BIOMEDICAL DEVICES AND THEIR APPLICATIONS

BIOMEDICAL APPLICATIONS OF LANTHANUM

BIOMEDICAL DEVICES AND THEIR APPLICATIONS

Additional books in this series can be found on Nova's website under the Series tab.

Additional E-books in this series can be found on Nova's website under the E-book tab.

BIOMEDICAL DEVICES AND THEIR APPLICATIONS

BIOMEDICAL APPLICATIONS OF LANTHANUM

HARRIET NILSSON
ANCA DRAGOMIR
AND
GODFRIED M. ROOMANS

Nova Biomedical Books
New York

Copyright © 2010 by Nova Science Publishers, Inc.

All rights reserved. No part of this book may be reproduced, stored in a retrieval system or transmitted in any form or by any means: electronic, electrostatic, magnetic, tape, mechanical photocopying, recording or otherwise without the written permission of the Publisher.

For permission to use material from this book please contact us:
Telephone 631-231-7269; Fax 631-231-8175
Web Site: http://www.novapublishers.com

NOTICE TO THE READER

The Publisher has taken reasonable care in the preparation of this book, but makes no expressed or implied warranty of any kind and assumes no responsibility for any errors or omissions. No liability is assumed for incidental or consequential damages in connection with or arising out of information contained in this book. The Publisher shall not be liable for any special, consequential, or exemplary damages resulting, in whole or in part, from the readers' use of, or reliance upon, this material. Any parts of this book based on government reports are so indicated and copyright is claimed for those parts to the extent applicable to compilations of such works.

Independent verification should be sought for any data, advice or recommendations contained in this book. In addition, no responsibility is assumed by the publisher for any injury and/or damage to persons or property arising from any methods, products, instructions, ideas or otherwise contained in this publication.

This publication is designed to provide accurate and authoritative information with regard to the subject matter covered herein. It is sold with the clear understanding that the Publisher is not engaged in rendering legal or any other professional services. If legal or any other expert assistance is required, the services of a competent person should be sought. FROM A DECLARATION OF PARTICIPANTS JOINTLY ADOPTED BY A COMMITTEE OF THE AMERICAN BAR ASSOCIATION AND A COMMITTEE OF PUBLISHERS.

Additional color graphics may be available in the e-book version of this book.

Library of Congress Cataloging-in-Publication Data

ISBN: 978-1-61728-906-4

Available upon request

Published by Nova Science Publishers, Inc. † New York

Contents

Preface		vii
Authors' Contact Information		ix
Chapter I	Lanthanum as a Tracer for the Study of Tight Junctions	1
Chapter II	Structure and Function of Tight Junctions	3
Chapter III	Methods to Study the Permeability of Tight Junctions in Cell Cultures	5
Chapter IV	Lanthanum in the Study of Tight Junctions in Different Tissues	11
Chapter V	Lanthanum as an Inhibitor of Ion Channels	23
Chapter VI	Uptake of Lanthanum Salts in the Body	27
Chapter VII	Lanthanum Carbonate in the Treatment of Hyperphosphatemia in End Stage Renal Disease	29
Acknowledgments		35
References		37
Index		55

Preface

Lanthanum has been used as a tracer to study tight junctions in epithelia by transmission electron microscopy. The key function of the tight junction complex (TJC) is to limit the paracellular permeability by forming a selectively permeable barrier between the apical and basolateral compartment of the extracellular space. Impaired regulation of the TJC causes a variety of diseases, such as autoimmune diseases, cancer development, infections, allergies and asthma. The advantage of lanthanum as an opaque probe is its small ionic radius (0.11 nm), compared to alternative methods, such as radioactive isotope techniques using extracellular tracers (e.g., [^{14}C] mannitol, molecular radius 0.39 nm) and light microscopy techniques with larger polar molecules (e.g., fluorescein isothiocyanate (FITC)-dextran). Other techniques to study tight junctions (TJ) are measurement of the transepithelial electrical resistance (TEER) and ionic conductance measurements. The lanthanum tracer technique was introduced by Revel and Karnovsky in 1967 and has been used in studies of the paracellular permeability in epithelia and endothelia in many different tissues, such as airways, blood brain barrier, testis, epididymis, intestine and liver. The drawback is that the tracer competes with calcium for the binding sites close to the TJ, which may cause artefacts. The lanthanum tracer method is sensitive and has been used to quantitatively assess acute and prolonged changes in the paracellular permeability. There is a good correlation between the lanthanum tracer method and the TEER method, which mainly reflects the paracellular ionic conductance. The fact that lanthanum competes with calcium is the basis for the use of La^{3+} ions as blockers of calcium (and other ion) channels, and the fact that La^{3+} ions bind phosphate ions is the basis for the use of lanthanum carbonate for the treatment of hyperphosphatemia in patients with end-stage renal disease.

Authors' Contact Information

Harriet Nilsson

Department of Medical Cell Biology, University of Uppsala,
Box 571, SE-75123 Uppsala, Sweden
Department of Biosciences and Nutrition, Karolinska Institutet,
Novum, SE-14157 Huddinge, Sweden

Anca Dragomir

Department of Medical Cell Biology, University of Uppsala,
Box 571, SE-75123 Uppsala, Sweden
Department of Clinical Pathology, Uppsala University Hospital,
SE-75185 Uppsala, Sweden

Godfried M. Roomans

School of Health and Medical Sciences, University of Örebro,
Örebro University Hospital, SE-70185 Örebro, Sweden

Chapter I

Lanthanum as a Tracer for the Study of Tight Junctions

The lanthanum tracer technique for the study of tight junctions (TJ) was introduced by Revel and Karnovsky (Revel and Karnovsky 1967). However, already earlier the use of lanthanum in electron microscopy was described by Lettvin and coworkers (Lettvin, Pickard et al. 1964) and Doggenweiler and Frenk (Doggenweiler and Frenk 1965) for the staining of membranes in nerve cells. La^{3+} had, by replacing Ca^{2+} ions, a stabilizing effect on membrane structure, but also an electron scattering power high enough to produce contrast in electron microscopic images. The lanthanum tracer technique was further modified by Shaklai and Tavassoli (Shaklai and Tavassoli 1977) to improve the detection of the tracer in the paracellular space through enhanced uniform precipitation.

The technique has ever since been widely used to assess the permeability of the junctional structures, especially the TJ in epithelia and endothelia (Todd, Inman et al. 2000). The tracer has been found to penetrate leaky epithelia, i.e., epithelia with a transepithelial electrical resistance (TEER) of <200 $\Omega \cdot cm^2$, such as gall bladder and ileum (Machen, Erlij et al. 1972) and renal tubules (Tisher and Yarger 1973), but not tight epithelia (with a TEER >1000 $\Omega \cdot cm^2$, such as urinary bladder and frog skin (Erlij and Martinez-Palomo 1972; Claude and Goodenough 1973; Todd, Inman et al. 2000). There may be a threshold TEER beyond which any further increase in resistance will not be accompanied by a significant decrease in permeability of marker molecules (Todd, Inman et al. 2000). The good staining and tracer properties in the electron microscope are assumed to largely depend on non-colloidal La^{3+} ions, since 80% of the lanthanum exists as charged particles of less than 500 daltons (Da) at pH 7.7 (Schatzki and Newsome 1975). Treatment with La^{3+} produces no significant alterations in cellular morphology (Sutiagin and Pylaev 1983).

Chapter II

Structure and Function of Tight Junctions

The key function of TJ is to limit the paracellular permeability and to form a selectively permeable barrier between the apical and basolateral compartment of the extracellular space (Mitic, Van Itallie et al. 2000). This specific barrier function can be characterized and quantified as TEER as well as charge selectivity (Colegio, Van Itallie et al. 2003). TJ also have a fence function restricting movement of membrane proteins, thus creating a distinction between ion channels, receptors and functionally specialized domains in the apical and basolateral parts of the plasma membrane, thus maintaining cell polarity (Gumbiner 1993; Mandel, Bacallao et al. 1993; Drubin and Nelson 1996; Kiessling, Kartenbeck et al. 1999). Impaired regulation of the TJ causes a variety of diseases coupled to abnormal paracellular permeability. In the embryo, some of these conditions are so serious that foetuses are not viable. Other examples of TJ-related diseases are autoimmune diseases, cancer development, infections, allergies and asthma (Cereijido, Contreras et al. 2007).

TJ are located in the most apical part of the cell and form a continuous belt-like region (100-800 nm deep) of intimate contact between the plasma membrane of adjacent cells (i.e. "membrane kisses"). The beltlike strand-containing regions act as a seal where the molecular characteristics and complexity of the network are correlated to TEER and thus the permeability of the TJ (Balda, Flores-Maldonado et al. 2000). There is a substantial physiological variation in the "tightness" of the TJ between different cell lines (Gruenert, Willems et al. 2004) and tissues (Claude and Goodenough 1973).

The scaffold of the junctions is formed by members of the membrane-associated guanylate kinase (MAGUK) homologue protein family (Dimitratos, Woods et al. 1999), namely the zonula occudens (ZO) proteins ZO-1, ZO-2, and ZO-3. Each of these scaffold proteins contains a sequence of protein-binding domains, including three PDZ-binding domains (i.e., post synaptic density protein 95, discs large, zonula occludens-1), an Src homology 3 (SH3) and a guanylate kinase (GUK) domain (Tsukita and Furuse 1999). ZO-1, ZO-2, and ZO-3 connect to the actin filaments, and in this way assemble different structural and regulatory proteins. Such proteins are kinases and G-proteins together with membrane proteins, which are essential for a proper function of the junction (Mehrotra, Martin et al. 2008). Activation of pathways modulating the phosphorylation pattern of cytoskeletal proteins associated with TJ are important for assembly and remodelling of junctional structures and this could change their permeability over time (Ivanov, McCall et al. 2006; Miyoshi and Takai 2008).

The membrane proteins are the junctional adhesion molecule (JAM), occludin and claudin (Furuse, Itoh et al. 1994; Itoh, Furuse et al. 1999; Wittchen, Haskins et al. 1999). Claudins and occludins form the strands of the TJ. The claudins have tissue specificity determining the barrier function (Tsukita and Furuse 1999; Tsukita, Furuse et al. 1999), whereas occludin has a regulatory function of the seal and the selective paracellular permeability (Balda, Flores-Maldonado et al. 2000; Huber, Balda et al. 2000; Ohtake, Maeno et al. 2003; Siu and Cheng 2004). (JAM)-1 is also important for regulation of the junctional integrity (Naik and Eckfeld 2003). It is involved in cellular adhesion (Yoshikumi, Ohno et al. 2008) and in the formation of cell polarity interacting with the polarity protein Par (partitioning-defective)-3 complex (Ebnet, Suzuki et al. 2001; Itoh, Sasaki et al. 2001). The formation and maintenance of TJ is also dependent on the cell-cell adhesion activity of E-cadherin, which involves Ca^{2+}. Ca^{2+} is essential to maintain intercellular contacts and is important for the stability of mature TJ (Lacaz-Vieira 1997). Antigens and mediators released by resident inflammatory cells as well as hyperosmolar solutions can alter the permeability of the TJ (Nusrat, Turner et al. 2000; Walsh, Hopkins et al. 2000).

Chapter III

Methods to Study the Permeability of Tight Junctions in Cell Cultures

There are different methods to measure transepithelial permeability in tissues and cell cultures, e.g., the lanthanum tracer technique, radio-isotope techniques using extracellular tracers and light microscopy techniques with large polar molecules, such as fluorescein isothiocyanate (FITC)-dextrans with an average molecular weight of at least 4 kD, or the use of electrophysiological methods such as TEER measurements.

The Lanthanum Tracer Method

A 4% stock solution of lanthanum nitrate pH 7.6-7.8 is prepared as follows: 4 g lanthanum nitrate was dissolved in 50 ml distilled water and the pH is adjusted to 7.6-7.8 with 0.1M NaOH while stirring vigorously and slowly adding the base (Revel and Karnovsky 1967). The volume is then brought to 100 ml with distilled water. One volume of this solution is added to 3 volumes of the fixative (e.g., 3.33% glutaraldehyde in 0.12 M sodium cacodylate buffer, pH 7.4) resulting in a final concentration of 2.5% glutaraldehyde and 1% lanthanum in 0.09 M cacodylate buffer. The fixative is made fresh immediately before use (Nilsson, Dragomir et al. 2006).

Cell cultures are kept in the fixative for at least 4 hours at room temperature, to prevent precipitation of lanthanum in the cold, and thereafter rinsed in a phosphate isoosmolar to the fixative, overnight at 4°C. The subsequent postfixation is performed in 1% OsO_4 in cacodylate buffer (0.1M, pH 7.4) for 30 min., containing 1% $La(NO_3)_3$ (to avoid washing out the tracer) (Shaklai and Tavassoli 1977; Nilsson, Dragomir et al. 2006). The cells are finally dehydrated in a graded ethanol series, embedded in epoxy resin and prepared for analysis in the transmission electron microscope.

For morphological experiments on monolayers of cultured cells, (Kiessling, Kartenbeck et al. 1999) suggested a method where the tracer was applied to both the apical and the basolateral side of a permeable support. However, to study whether a particular treatment increases the permeability of the TJ, this morphological method is less suitable, since the tracer penetrates the entire paracellular space from the basal side independently of the tightness of the TJ. It is hence better to apply the tracer only from the apical side. Nilsson and coworkers (Nilsson, Dragomir et al. 2007) used two different buffers, one added to the apical and one to the basal side of the cells. To the apical buffer 1% $La(NO_3)_3$ was added as a TJ permeability marker (the $La(NO_3)_3$ being included in the total osmolarity of the buffer. The basal buffer did not contain $La(NO_3)_3$, but NaCl or xylitol was added to the "basal" buffer to compensate for the osmolarity of lanthanum in the "apical" buffer.

The opening state of the TJ was semi-quantitatively expressed according to a scheme proposed by (Högman, Mork et al. 2002), in which the TJ were classified as one of three types: 1) no penetration of lanthanum tracer at all, "intact TJ" 2) penetration of lanthanum for a short distance in the TJ, "weakened TJ" and 3) penetration of the lanthanum through the entire TJ into the paracellular space, "open TJ" (Figure 1a,b).

The advantage of lanthanum as a tracer is that its dimensions are similar to those of small ions of physiological interest (Schatzki and Newsome 1975; Todd, Inman et al. 2000). The major determinant of lanthanum penetration is likely to be the complexity and tightness of the TJ rather than anionic fixed charges of the structure (Todd, Inman et al. 2000). The drawback of La^{3+} is that the tracer competes with Ca^{2+} for Ca^{2+}-binding sites on TJ, due to its trivalency and an ionic radius (0.102 nm) that is about equal to that of Ca^{2+} (de Jong, Bosch et al. 1976). This may cause an opening of the TJ before the action of the fixative has taken place (Lacaz-Vieira 1997). However, at least in cell cultures where the fixation time is very short (a few minutes) this effect may be of minor importance.

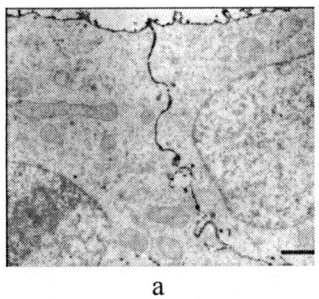

a

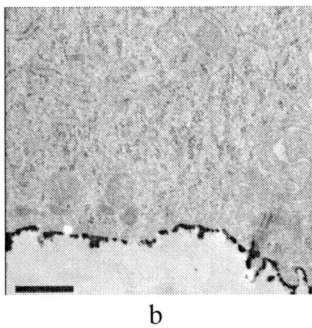

b

Figure 1. (a) Ultrastructural studies on cells exposed to 450 mOsm NaCl. A precipitation of lanthanum in the paracellular space is seen in cells with otherways intact morphology. The bar corresponds to 0.6 μm. (b) Ultrastructural studies at isosmotic conditions. Precipitation of lanthanum is seen apically of the cells. There is no penetration of lanthanum in the lateral intracellular space and the overall morphology is intact. The bar corresponds to 0.5 μm.

Transepithelial Electrical Resistance

Tight junction integrity in cell cultures grown on clear, permeable filter inserts (Costar Transwell®, Corning, Acton, MA, USA, 6.5 mm diam., 0.4 μm pore size) was assessed by measuring TEER with an EVOM epithelial voltmeter in an ENDOHM-6 chamber (World Precision Instruments, Sarasota, FL, USA). The sensitive parameters in this method, which have to be kept constant throughout the whole experiment, are temperature, pH and volume of the upper and lower reservoirs of the chamber. In the studies by Nilsson and cooworkers (Nilsson, Dragomir et al. 2006; Nilsson, Dragomir et al. 2007;

Nilsson, Dragomir et al. 2009), the temperature was kept at 37°C in the ENDOHM chamber and the cultures were left at this temperature during all preparation and experimental steps. To be able to measure TEER, the volume in the upper and lower reservoirs of the filter inserts was changed from 100 and 700 µl, to 200 and 1200 µl, respectively. The cells were allowed to equilibrate in the new volumes for 30 minutes before the experiment. As the resistance is inversely proportional to the area of the tissue, the product of the resistance and the area is calculated, (i.e., TEER). The TEER value, given as $\Omega \cdot cm^2$, mainly reflects the open-closed state of the TJ (Mehrotra, Martin et al. 2008). To determine background levels, TEER was measured, as described above, on the clear permeable filter inserts coated with 0.01% bovine collagen type I or with fibronectin solution, i.e., the same coating solution used when growing the cells. Measurements were performed every tenth minute and repeated until a stable value was obtained (after 30-60 minutes). The background of the clear permeable filter inserts without cells gave a stable TEER of ~20 $\Omega \cdot cm^2$ in all experiments.

With luminally added lanthanum there was a strong negative linear relationship between TEER and the paracellular permeability ($r^2 > 0.8$) (Nilsson, Dragomir et al. 2006; Nilsson, Dragomir et al. 2007; Nilsson, Dragomir et al. 2009). Thus, a decrease in TEER was paralleled by an increase in permeability and vice versa. This was seen both in experiments with different osmolarities (Nilsson, Dragomir et al. 2006; Nilsson, Dragomir et al. 2007) and in airway cultures exposed to inhibitors (CFTR$_{inh}$-172) and stimulators of conductance (IBMX and forskoline) (Nilsson, Dragomir et al. 2009).

Hydrophilic Permeability Probes

Radionuclide uptake in and efflux from isolated cells can detect enzyme activity, transport, and permeability of the apical and basolateral membranes. The weakness of this method is that it cannot differentiate a carrier from a channel (Sun and Gondos 1986), nor identify the exact permeability route (Todd, Inman et al. 2000). To circumvent this problem hydrophilic permeability probes such as [^{51}Cr]-EDTA or [^{14}C]-mannitol can be used, which are markers that permeate mainly through the paracellular pathway (Bjarnason, MacPherson et al. 1995; Nylander, Pihl et al. 2003). The transepithelial transport of [^{14}C]-mannitol across the epithelial cell monolayer was assessed by the apparent permeability coefficient, P_{app}, according to the following equation:

$$P_{app} = \frac{1}{AC_0}\frac{dQ}{dt},$$

where Q is the amount of [^{14}C]-mannitol transported into the basolateral chamber, t is the elapsed time, A is the area available to transport, and C_0 is the initial concentration of [^{14}C] mannitol. The relatively large radius of 0.47 nm of ^{51}Cr-labeled EDTA (Bustamante and Setchell 2000) or 0.34 nm for mannitol (Seki, Harada et al. 2008) make these molecules suitable for studies of leaky epithelia, but they may be less sensitive for the study of tight epithelia.

The intercellular penetration of FITC-dextrans is approximately size-dependent. These molecules are mainly used in penetration experiments in leaky epithelia or skin (Braga, Desterro et al. 2004). For a molecule of 4 kDa the radius is calculated to about 2.3 nm (Dogic, Purdy et al. 2004).

In general, the tracer techniques are slow since they are based on the principle of diffusion. This makes them less sensitive to the open-closed state of the TJs compared to the TEER measurements, since TEER is an instantaneous measurement (Mehrotra, Martin et al. 2008).

As mentioned above, we found that the linear relationship between TEER and the paracellular permeability to lanthanum is strong whereas this is not always the case when comparing TEER and P_{app}. It is important to keep in mind, that the two parameters, TEER and P_{app}, do not measure the same properties of TJ permeability. TEER measures predominantly the ionic conductance of the paracellular pathway, which is specified by the profile of claudins expressed, while the flux of non-electrolyte tracers (expressed as P_{app} or cumulative fraction transported) measures the paracellular water flow and may be proportional to the number of pores in the junctional structure (Ranaldi, Consalvo et al. 2003; Van Itallie, Holmes et al. 2008). It has been shown that diverse pathways, corresponding to various pore size and charge-selectivity in the TJ structure, could be differentially regulated (Watson, Hoare et al. 2005).

Chapter IV

Lanthanum in the Study of Tight Junctions in Different Tissues

Brain and Peripheral Nervous System

The blood-brain barrier (BBB) is an important functional unit of the nervous system, characterized by the special transport permitted between the capillaries and the nervous tissue, which allows brain homeostasis. The barrier consists of the endothelial cells of the capillaries, a thin basal lamina, and the glial cell feet (astrocytes, ependymal cells) that enclose the capillaries. Most of the capillaries of the nervous system are of the non-fenestrated type, with belt-like zonulae occludentes consisting of highly selective tight-junctions allowing diffusion of small lipofilic molecules (O_2, CO_2, hormones), and preventing diffusion of larger molecules or hydrophilic substances. The endothelial cells and the astrocytes have specific transport systems, both active and carrier-mediated, which support and regulate entry of ions, glucose, amino acids, and other metabolites. In addition, transport of large molecules is permitted by trans-endothelial vesicle trafficking (pinocytosis), although this is much less active than in non-cerebral capillaries.

Under physiological conditions, lanthanum does not penetrate the BBB and is exclusively located in the lumen of the capillary. Changes in the BBB lead to lanthanum leakage from the capillaries and staining of the perivascular parenchyma, in addition to morphological changes such as vacuolation, membrane damage, and cell edema. This observation is of value for studies of brain injury and repair, and for testing of new therapeutic agents. Lanthanum was e.g., used early to demonstrate that the BBB is reversibly opened by hyperosmotic mannitol (Nagy, Pappius et al. 1979), hypertension (Nagy, Mathieson et al. 1979), hypothermia (Schindelmeiser et al., 1987), acute ischemia (Nakagawa et al., 1990), and venom (da Silva et al., 2004).

Nag and coworkers(Nag, Robertson et al. 1982) showed that under physiological conditions, rat arterioles in the brain have actually leaky cell-cell contacts, since lanthanum was found in the arteriolar wall and in the extracellular compartment of the surrounding brain. This supports the notion that TJ in the arteriolar endothelium are not forming a continuous circumferential band like those in the capillaries. Acute hypertension resulted in faster tracer passage into the brain tissue, through not only the arteriolar walls but also through capillaries and venules.

Lanthanum was used to demonstrate that astrogaloside IV, a small molecule extracted from a traditionally used medicinal herb (*Astragalus membranaceus*), maintains the BBB after ischemic/reperfusion injury in a murine model (Qu, Li et al. 2009). The mechanism for this is supposedly an upregulation of the expression of the TJ proteins occludin and ZO-1 and inhibition of the release of inflammatory cytokines from endothelial cells.

Lanthanum tracer demonstrated the development of a BBB-like system in invertebrates, with a period of incomplete tightens in hatchlings (Leslie 1975). The morphological changes in lanthanum permeability coincided with the onset of mature neuronal electric proprieties and spontaneous synaptic activity (Swales and Lane, 1985). Repair of injured glial cells can be followed with a lanthanum tracer (Smith et al., 1984), showing in the nervous system of invertebrates that morphological observations are tightly correlated with re-establishment of the normal permeability properties of BBB.

In rats, the TJ of BBB restricted lanthanum passage already in 1 day-old animals, although the basal lamina is ill-defined during the first 14 days (Xu and Ling 1994). In humans it was noted that well-formed TJ are present in the brain capillaries and choroid plexus cells as soon as these cells differentiate in embryos and fetuses. Thus, the very high concentration of proteins in the fetal CFS compared to an adult cannot be accounted for by a lack of TJ in the developing brain (Mollgard and Saunders 1986).

In a model of immune-mediated de-myelination (chronic relapsing experimental allergic encephalomyelitis) it was noted that the BBB was disrupted not by altered TJ but by increased vesicular transport, which was associated with a metabolic change in the endothelial cells (Hawkins, Munro et al. 1992). In experimental obstructive hydrocephalus, lanthanum tracer indicated that the TJ between ependymal cells in the ventricles allow paracellular transport of small solutes while restricting the movement of larger molecules (Nakagawa, Cervos-Navarro et al. 1985).

There are several places in the nervous system where the BBB is weakened by the presence of fenestrated capillaries, termed circumventricular organs. This allows secretion of neurohormones directly into the blood stream (i.e., the pineal gland, the posterior hypophysis gland) and monitoring of blood composition in specialised brain parts (e.g., in the area postrema, which triggers vomiting when the blood contains toxic substances, in the lamina terminalis, which contains chemosensors, and in the subfornical organ which senses body fluids) (Ganong 2000).

Anatomical subdivisions of the BBB are the blood-cerebrospinal fluid barrier (B-CFS-B) and the blood-ocular barrier, having a similar role in maintaining the homeostasis of the special environments. The B-CFS-B is found in the choroid plexus, formed by endothelial cells, basal membrane, and choroid plexus cells. These ependymal cells are cuboidal/cylindrical, with microvilli for absorbing cerebrospinal fluid (CFS), and cilia for circulating the CFS. The choroid plexus cells are connected by TJ, and produce CFS from the capillaries through filtration. The capillaries of choroid plexus are fenestrated, and have gaps between endothelial cells. Hence the TJ between the cells of the choroid plexus are the site where the diffusion of molecules over the B-CFS-B can be regulated. However, the total permeability in the B-CSF-B is lower than that of the BBB. Special proteins can transport molecules (e.g., thymidine) through the B-CSF-B that are not present at the BBB (Wu, Yuan et al. 1992). Lanthanum tracer studies showed that in rats, the area postrema allows direct communication between blood and ventricular cerebrospinal fluid (CSF) through an incomplete junction of the ependymal cells, which represents a transitional form between TJ and adherens junctions (Gotow and Hashimoto 1979). The permeability of B-CFS-B is increased by atrial natriuretic factor, as noted by the passage of both lanthanum and horseradish peroxidase from blood into the choroid plexus through the apical TJ and the choroidal epithelial cells (Nag 1991).

In contrast to the central nervous system, the periferal nervous system lacks a blood-tissue barrier, since the cells of endoneurial capillaries are connected by TJ which are weaker than those in the brain vessels, which allows lanthanum passage, findings which are compatible with data from physiological experiments (MacKenzie, Ghabriel et al. 1987). The paracellular transport is restricted by the presence of the ZO within the perineurium. However, the perineurium is open-ended at some nerve terminations, at least with regard to ionic diffusion. In invertebrates the glial cells building the perineurium are also interconnected by leaky TJ, as observed by lanthanum tracer technique (Lane, Harrison et al. 1981).

Eye

The blood-ocular barrier (BOB) is formed by non-fenestrated capillaries, the basal membrane and the retinal/ciliary body epithelial cells. The endothelia of the retina have complex belt-like ZO consisting of highly selective TJ. The adjacent retinal pigmented epithelial cells are also sealed by TJ, restricting the permeability even further. Okinami and coworkers (Okinami, Ohkuma et al. 1976) studied the characteristics of this barrier in albino rats. They found that there are several layers that are impermeable to lanthanum tracer, due to cells adjoined by TJ: (1) the endothelial cells of the retinal capillaries, (2) the retinal pigment epithelial cells, (3) the glial cells of Kuhnt, intermediary cells, surrounding the cone and rod cells, and (4) the cells of the outer limiting membrane. Furthermore, the cells of the inner limiting membrane were also impermeable to lanthanum, despite their lack of a ZO, and a functional barrier mechanism was suggested for this layer.

Similar to the BBB, the BOB can be disrupted by opening TJ under hyperosmotic stress (Okinami, Ohkuma et al. 1978), prostaglandin E1 (Pedersen 1980), degenerative retinal diseases such as inherited retinal degeneration (Caldwell and McLaughlin 1983), or diabetes (Caldwell, Slapnick et al. 1985), and ischemia (Kaur, Foulds et al. 2008).

The lens presents in the periphery, at least in the eye of frog and rat, superficial epithelial cells adjoined by TJ impermeable to lanthanum and horseradish peroxides (HRP) tracer, while the equatorial region is permeable and represents the major site of efflux and influx (Lo 1987).

Gastrointestinal Tract

The gastrointestinal tract is along its entire length covered by epithelium. Starting with the teeth, TJ have been demonstrated in the epithelial lining of the tract by use of lanthanum tracers in electron microscopy. Bishop (Bishop

1985) used the permeability of fenestrated capillaries amongst the odontoblasts to deposit lanthanum in the interodontoblastic space. Towards the predentine, often less than 10 µm from the capillaries, the progress of the tracer was abruptly halted by the TJ at the apical ends of the odontoblast cell bodies, so that lanthanum was not found in the predentine. This means that all substances deposited in predentine and dentine must arrive by passing through the odontoblasts.

TJ have also been demonstrated by the lanthanum technique in the palatal epithelium (Martin, Appleton et al. 1987). These TJ could be made permeable if the yeast *Candida albicans* was inoculated between an acrylic plate prosthesis and the palatal epithelium; this was not the case if the palate was exposed to the yeast in the absence of the prosthesis. Similarly, the presence of TJ was shown in the uppermost layer of the stratum granulosum in the lingual epithelium (Holland, Zampighi et al. 1989).

In the esophagus, TJ were demonstrated in series within stratum corneum; less commonly, they extended to two orthree cell layers of upper stratum spinosum; however, this study did not find support for a major role for these junctions in barrier function in this tissue (Orlando, Lacy et al. 1992).

In the epithelium of the stomach, it was demonstrated that aspirin treatment resulted in alterations in tight junction complex morphology and an increase in permeability to the lanthanum tracer, which suggests that aspirin-induced impairment of the tight junction complexes between gastric mucosal epithelial cells may be a major contributing factor in the etiology of stomach disorders (Meyer, McGinley et al. 1986).

The lanthanum technique was used to study the epithelial permeability in the rat small intestine (Madara and Trier 1982). Dense lanthanum precipitates in TJ and paracellular spaces were restricted to a subpopulation of villous goblet cells and were not found between villous absorptive cells. These TJ were also permeable to barium, but not to macromolecular tracers such as microperoxidase, cytochrome c and horseradish peroxidase. It was also shown that palmitoylcarnitine (PCC) opens TJ in a monolayer of Caco-2 colon carcinoma cells; this phenomenon appears to be responsible for the significant enhancement of the absorption of hydrophilic drugs across intestinal mucosa caused by PCC and other long-chain acylcarnitines (Hochman, Fix et al. 1994). In an experiment on rats, it was demonstrated that immobilisation stress induced a significant (but reversible) increase in epithelial permeability to the lanthanum tracer (Mazzon, Sturniolo et al. 2002).

Small intestine permeability is frequently altered in patients with Crohn's disease, and this has been the subject of several studies in which the lanthanum tracer technique was used. In experimentally induced colitis (with trinitrobenzenesulfonic acid/ethanol) in rats, 80% of duodenal and 73% of ileal TJ were permeable to lanthanum, whereas in controls, only 8% of duodenal and 10% of ileal junctions were leaky. In colitis, the percentage of "leaky" junctions in the duodenum and the terminal ileum correlated positively with the macroscopic colon damage score (Fries, Mazzon et al. 1999). In another (mouse) model for colitis (induced by dinitrobenzene sulfuric acid, DNBS) an increase of tight junction permeability throughout the entire small intestine was found and again, the extent of alterations in permeability was correlated with colonic damage (Mazzon and Cuzzocrea 2007). In this animal model, endogenous peroxisome proliferator-activated receptor-α (PPAR-α) ligands reduced small intestinal permeability through the regulation of tight junction protein and apoptosis. High dose zinc treatment may give protection against colitis-induced alterations in tight junction permeability (Sturniolo, Fries et al. 2002). Experimental colitis in rats was associated with increased serum bilirubin and bile acid concentrations, a 2.5-fold increase in paracellular biliary excretion of horseradish peroxidase, and a nine-fold increase in lanthanum permeability through the hepatocyte TJ, although liver histology, cingulin and ZO-1 localizations were similar to normal liver, indicating that subtle alterations in tight junction function may be involved in the pathogenesis of hepatobiliary injuries in inflammatory bowel disease (Lora, Mazzon et al. 1997).

In a study of guinea-pig cecal epithelium, Mora-Galindo (Mora-Galindo 1986) showed that cells at the surface and in middle regions of crypts possess TJ impermeable to lanthanum, whereas junctions between cells located at the bottom of crypts often were permeable to the tracer, indicating that permeability decreases as the epithelial cells mature.

In a series of studies on liver, it was found that treatment with the antimicrotubular drugs vinchristine or colchicine results in dilated intercellular spaces and passage of lanthanum in the canalicular lumen, i.e., the interhepatocyte TJ became permeable to the tracer; this was taken to indicate that the intact microtubular system, possibly via microfilaments, may play a role in the formation of TJ (Rassat, Robenek et al. 1981; Rassat, Robenek et al. 1982). In cholestasis, experimentally induced in rats by long-term treatment with estradiol valerate, lanthanum did not penetrate the TJ, indicating that in the liver one tight junctional strand is sufficient to prevent the escape of larger bile constituents such as bile acids, and that a back diffusion of bile acids over the tight junctional barrier does not play a role in the pathogenesis of the estrogen-induced cholestasis (Robenek, Rassat et al. 1982).

Sometimes, hyperbilirubinemia without mechanical obstruction of the biliary tree is encountered postoperatively; patients with this complication often suffer from bacterial infections and endotoxemia, which results in stimulation of Kupffer cells to secrete tumor necrosis factor-α (TNF-α) and interleukin (IL)-6. In a study by Ikeda and coworkers (Ikeda, Mitaka et al. 2003), bile was collected from rats treated with or without these cytokines. The livers, perfused with lanthanum after the injection of the cytokines, were examined ultrastructurally. In rats treated with cytokines, the total serum bile acid concentration increased and lanthanum temporarily accumulated in the bile canaliculi, which suggests that the cytokines may reduce bile canaliculi contractions and thereby decrease bile flow.

Respiratory System

Net vectorial fluid transport in human airway epithelium depends critically on ENaC (epithelial sodium channel) and the cyclic adenosine monophosphate (cAMP)-activated chloride channel in the apical membrane, the cystic fibrosis transmembrane regulator (CFTR) operating in concert with the paracellular and transcellular pathways. In some animals, Ca^{2+}-dependent chloride channels are also present and may be more important than CFTR (Alton, Manning et al. 1991). Fluid absorption is mainly controlled by the transport of Na^+ through ENaC, while fluid secretion is regulated by cell to lumen movement of Cl^- and/or HCO_3^- via CFTR (Blaug, Hybiske et al. 2001). In both cases, the obligatory movement of counterions likely takes place predominantly through leaky TJ (Blaug, Hybiske et al. 2001). The resulting fluid dilutes the airway mucus and contributes to the fluid bathing the cilia. The composition of this airway surface liquid (ASL) has long been a matter of debate, but most data indicate that it is normally close to isosmotic (Hull, Skinner et al. 1998; Kozlova, Vanthanouvong et al. 2005; Kozlova, Vanthanouvong et al. 2006).

Inhalation of hyperosmotic solutions (of e.g., salt, mannitol) has been used in the treatment of patients with cystic fibrosis, bronchiectasis or asthma (Daviskas, Anderson et al. 1996; Daviskas, Anderson et al. 1997; Daviskas, Robinson et al. 2002; Wark, McDonald et al. 2005; Donaldson, Bennett et al. 2006; Elkins, Robinson et al. 2006). The effect of high concentrations of NaCl or mannitol in the airways likely involve a number of factors, and changes in the permeability of the paracellular pathway may be one of these (Relova and

Roomans 2001; Högman, Mork et al. 2002; Nilsson, Dragomir et al. 2007). This may result in an increased transepithelial water transport mainly via the paracellular route followed by an increased ASL volume with more diluted mucus, which would be easier for the patient to eliminate. Transiently raised osmolarity via aerosol administration with a subsequently induced water flux towards the luminal surface of the epithelial cells (secretion) has been described (Hirsh 2002). An increase in mucus transport associated with an increase in water has also been shown in animal studies in vivo (Winters and Yeates 1997; Winters and Yeates 1997). Studies where the lanthanum tracer was added to the luminal side of the epithelium showed that the electron dense lanthanum could be visualised by electron microscopy in the lateral intercellular spaces, where it increased with increased osmolarity. Hence, hypertonic fluids opened the TJ, which confirmed the hypothesis mentioned above (Högman, Mork et al. 2002; Nilsson, Dragomir et al. 2007).

Thymus

Raviola and Karnovsky (Raviola and Karnovsky 1972) demonstrated the existence of a blood-thymus barrier with the help of, among other techniques, lanthanum tracer. Morphological studies showed that most thymic capillaries were of the non-fenestrated type, with endothelial cells adjoined by impermeable TJ, lying on a basal lamina. The capillaries were surrounded by reticular cell projections and macrophages, which, alone, did not prevent tracer passage but had a great capacity of capturing foreign molecules. A small number of fenestrated capillaries was found in the extreme periphery of the thymic cortex, usually close to connective tissue. For the medulla, free transfer of macromolecules from venules was observed, which allowed lymphocyte migration. Thus the concept of a blood-tissue barrier that allows maturation of cortical lymphocytes in an immune-isolated compartment was accepted only for the cortex (Raviola and Karnovsky 1972).

Male Reproductive Organs

Testis

In the seminiferous tubules of the testis, the spermatozoa develop, starting as diploid spermatogonia and ending as haploid spermatozoa. The haploid spermatozoa would be recognized as "foreign" by the body's immune system. To avoid an autoimmune reaction to the developing spermatozoa, there is a physical barrier between the diploid and the haploid stages of the developing spermatozoa. This physical barrier, maintained by the endothelial cells of the non-fenestrated capillaries, a continuous basal membrane, and the Sertoli epithelial cells, connected by TJ, is called the "blood-testis barrier". However, the TJ and adherens junctions between the Sertoli cells must open periodically to accommodate the migration of spermatogonia into the lumen of the tubuli.

With the lanthanum tracer technique, the TJ between the Sertoli cells have been investigated in various mammals, such as the rabbit (Sun and Gondos 1986) the mink (Pelletier 1986), the fowl (Bergmann and Schindelmeiser 1987), the rat (Cavicchia and Sacerdote 1988; Cavicchia and Sacerdote 1991), the dog (Cambrosio Mann, Friess et al. 2003), and the human (Bergmann, Nashan et al. 1989). Development of TJ impermeable to lanthanum tracer coincides with the spermatocytes reaching or passing the zygotene-pachytene stages of meiosis I (Cavicchia and Sacerdote 1991; Morales, Mohamed et al. 2007), which occurs at puberty. Before the TJ are formed, the developing spermatozoa are prone to undergo apoptosis, presumably because they are left exposed in an open environment instead of being isolated in the adluminal compartment to which they are destined (Morales, Mohamed et al. 2007). The basement membrane in the testis (a modified form of extracellular matrix) is important in germ cell movement across the TJ between Sertoli cells because proteins in the extracellular matrix were shown to regulate these junctions via interactions between collagens, proteases, and protease inhibitors, possibly under the regulation of cytokines (Siu and Cheng 2004).

The blood-testis barrier is compromised by vitamin A deficiency (Huang, Yang et al. 1988; Morales and Cavicchia 2002), cytochalasin D (an actin filament disrupting substance) (Weber, Turner et al. 1988), chromium (as potassium dichromate) (Murthy, Saxena et al. 1991) cadmium (Yang, Schryvers et al. 2006), electromagnetic pulses (Wang, Ding et al. 2008), as well as during irregular hypospermatogenesis (Meyer, Mezrahid et al. 1996) and aging (Levy, Serre et al. 1999). The blood-testis barrier is, however, functionally normal in the intra-abdominal testis of cryptorchid rats (Hatier and Grignon 1986) or in the testis of aspermic (mutant) rats (Atagi, Ikadai et al. 1993).

Epidydymis

The epididymis is actively involved in the sperm maturation process (acquisition of motility and fertilizing ability), not only providing an appropriate environment but also providing many of the molecules needed by the spermatozoa to allow them to fertilize an egg. Tight junctional complexes between the epididymal cells form what is referred to as the blood-epididymis barrier (BEB), an important physiological and anatomical barrier that allows the epididymis to create a specialised fluid environment for the maturing spermatozoa. Another function of the BEB is to convey immunological protection of the spermatozoa (similar to the blood-testis barrier). It is the fluid microenvironment within the epididymis that has been suggested to promote maturation of the spermatozoa. The fluid is hyperosmotic and distinctly different in composition from blood plasma. The BEB regulates the entry of solutes and ions into the lumen, and the luminal fluid contains antioxidants, e.g., glutathione, and enzymes such as γ-glutamyl transpeptidase, superoxide dismutase and glutathione-S-transferase involved in antioxidant defence and protection against xenobiotics. Lanthanum has been used for studying the development of BEB in immature rats (Agarwal and Hoffer 1989). The epididymis and testis of the stallion were studied by López and coworkers (Lopez, Fuentes et al. 1997). Striking differences were observed in the penetration of lanthanum tracer and hence in the geometrical organization of the TJ along the ductus efferentes, epididymides and vas deferens. The flow of tracer was not impeded by the vascular endothelium. The TJ of the ductuli efferentes are poorly developed but unlike those of rats, guinea pigs or man they are not associated with gap junctions. The barrier of the ductuli efferentes corresponds to the 'leaky type'. In the epididymis the TJ are well developed throughout the duct, especially in the cauda epididymidis regions. Also Guan and coworkers (Guan, Inai et al. 2005) studied the permeability of epididymal TJ, in relation to the variance by segment of the luminal fluid microenvironment. TJ strands were continuous and impermeable to lanthanum nitrate on postnatal day 7, suggesting the establishment of functional TJ.

Female Reproductive Organs

Winterhager and Kühnel (Winterhager and Kuhnel 1985) studied the permeability barriers of the multilayered vaginal epithelium using lanthanum tracer techniques. During diestrus and proestrus the upper layers of mucified epithelial cells exhibit tight-junctional belts. When these cells begin to degenerate toward the end of proestrus the underlying epithelium is already keratinized as typical for estrus. The keratinized epithelial cells have a tight-junctional network that joins the basal plasma membranes with the apical membranes of subjacent cells and blocks paracellular diffusion of the tracer molecules. During conversion of the cornified epithelium to a mucified epithelium in metestrus the intercellular space of the epithelium is stained by tracer molecules even though tight-junctional belts can be observed. It can be concluded that in diestrus and proestrus, the TJ restrict paracellular diffusion; in metestrus, however, the TJ become functionally leaky although they remain morphologically intact.

The placental barrier in rats can be disturbed by fetal hypertension that opens paracellular pathways through the syncytial layer of the trophoblast (Kertschanska, Stulcova et al. 2000). On the other hand, Hua and coworkers (Hua, Zhu et al. 2009) showed that contrast-enhanced ultrasound does not increase the permeability of placenta to macromolecules larger than lanthanum or albumine.

Other Tissues

Skin

Neutral lipid storage disease with ichthyosis (NLSDI; Chanarin-Dorfman syndrome) is an ichthyosiform syndrome. Basal permeability barrier function and stratum corneum (SC) integrity were found to be abnormal (Demerjian, Crumrine et al. 2006). The basal barrier abnormality was linked to the secretion of lipid micro-inclusions, first segregated within lamellar bodies, which then form a non-lamellar phase within the SC interstices. With colloidal lanthanum nitrate perfusion, excess water/solute movement was restricted to the SC interstices, and further localized to non-lamellar domains. Phase separation of excess stored lipid provides a unifying pathogenic mechanism not only for NLSDI, but also in several other inherited ichthyosiform disorders of lipid metabolism, such as recessive X-linked ichthyosis and type 2 Gaucher's disease. Jiang and coworkers (Jiang, Chu et al. 2007) assessed the effects of ultraviolet B (UVB) irradiation in epidermal barrier function in murine epidermis. Electron microscopic observations demonstrated that the water-soluble lanthanum tracer was present in the extracellular SC domains, and the increased intercellular permeability was correlated with defective organization of the extracellular lipid lamellar bilayers of the SC.

Urinary Bladder

Kreft and coworkers (Kreft, Romih et al. 2002) established an *in vitro* culture model that closely resembles whole mouse urothelial tissue. When examined at the ultrastructural level, the cultured urothelium was polarized and organized as a multilayered epithelium, and epithelial organization was stabilized by well developed cell junctions. Desquamation of urinary bladder epithelial cells can be induced by constant illumination for 96 hours or by application of endotoxin lipopolysaccharide. Cell detachment involves interruption of TJ between neighbouring cells (Jezernik 1996). Urothelium permeability was studied in an experimental ischemic model of mouse urinary bladder by means of lanthanum nitrate tracer (Korosec and Jezernik 2000). Ischemia induces breakdown of the blood-urine permeability barrier by disruption of the TJ.

Nonsteroidal anti-inflammatory drug-induced cystitis is caused by indomethacin. Oral indomethacin treatment in mice caused penetration of lanthanum nitrate through intercellular areas of the epithelium, as well as an excess of mast cells. Hence, indomethacin resulted in histopathologic findings typical of interstitial cystitis, such as leaky bladder epithelium and mucosal mastocytosis (Cetinel, Cetinel et al. 2003).

Chapter V

Lanthanum as an Inhibitor of Ion Channels

Due to the fact that it resembles Ca^{2+}, lanthanum may compete with Ca^{2+}, replace or displace Ca^{2+}, and block Ca^{2+} uptake or efflux by cells and organelles; hence the effects of ionic lanthanum on different cells and organs are complex.

Heart

Lanthanum causes redistribution of Ca^{2+} ions in cardiac muscle cells, suppressing the zero-sodium response (e.g., a maintained contracture, and an asynchronous localized contraction) (Mead and Clusin 1985). Lanthanum also blocks the inward Ca^{2+} current in guinea pig (Wendt-Gallitelli and Isenberg 1985; Ravens, Steinmann et al. 1987) and bullfrog (Nathan, Kanai et al. 1988) heart muscle cells, thereby blocking contractility. In addition, lanthanum inhibits the ATP-dependent component of Ca^{2+}-efflux (Brommundt and Kavaler 1987).

Nervous System

Voltage-gated calcium currents (VGCCs) evoked by depolarizing voltage steps are reversibly blocked by lanthanum in neurons from the dorsal horn of the rat spinal cord but lanthanum enhances kainate-evoked responses at the same concentrations at which it suppresses VGCCs (Reichling and MacDermott 1991). Lanthanum inhibits ^{3}H-nitrendipine binding to brain membranes and blocks the stimulating actions of Ca^{2+} (Gould, Murphy et al. 1982), and it inhibits light-induced Ca^{2+} influx into *Drosophila* photoreceptors (Rom-Glas, Sandler et al. 1992). Blocking of Ca^{2+} fluxes by lanthanum may have consequences for the behavior of experimental animals. Lanthanum administered immediately after a visual reminder presented to day-old chickens following a single trial passive avoidance learning task produced an immediate but transient loss of memory on retention test. Summers and coworkers (Summers, Crowe et al. 1996) and Che and coworkers (Che, Cui et al. 2009) recently showed that lanthanum causes significantly impaired long-term memory in chickens. In view of the fact that lanthanum presently is used as a drug to treat hyperphosphatemia in humans (see below), these possible neurotoxic effects of lanthanum should not be neglected.

Neuromuscular Synapses

Lanthanum can act as a surrogate for Ca^{2+} in transmitter release at (mouse) motor nerve terminals, and cause an increased frequency of miniature end-plate potentials at the neuromuscular junction (Curtis, Quastel et al. 1986). According to Provan and Miyamoto (Provan and Miyamoto 1992) lanthanum enters the terminal through Na^+ channels and promotes Ca^{2+} release from intracellular organelles.

Endocrine System

Lanthanum binds to the cell surface of parathyroid cells (without entering the cells), but this gives rise to increased Ca^{2+} permeability resulting in a rise in intracellular Ca^{2+} (Gylfe, Larsson et al. 1986). Parathyroid hormone (PTH) release from the gland is normally inhibited by Ca^{2+}, but is inhibited by lanthanum to an even greater extent. Incubation of parathyroid cells with pertussis toxin, a G-protein inactivator that blocks inhibition by Ca^{2+}, does not block the inhibition of PTH release by lanthanum. There may be two cell surface sites that recognize lanthanum and Ca^{2+} independently (Fitzpatrick 1990).

Lanthanum acts in a Ca-like manner in the regulation of the cytosolic free Ca^{2+} concentration and in the secretion in melanotrophs in the rat pituitary gland (Shibuya and Douglas, 1992). Pb^{2+} entry in pituitary cells was inhibited by lanthanum, due to its properties as a blocker of capacitative Ca^{2+} entry (Kerper and Hinkle 1997).

Lanthanum can enter (bovine) chromaffin cells via the Na/Ca antiporter independently of, or together with Ca^{2+}, but, high concentrations of La^{3+} block the influx or efflux of Ca^{2+}. La^{3+} is at least as effective as Ca^{2+} in triggering catecholamine release and maintaining prolonged release, and acts cooperatively with Ca^{2+} at the release pathway (Powis, Clark et al. 1994).

Male Reproductive System

Lanthanum abolished both the tonic and the phasic KCl-induced contractions in the rat vas deferens, and this was not reversed by Ca^{2+} (Hay and Wadsworth 1982). Lanthanum reduced the Ca^{2+}-dependent contractions induced by oxytocin in the isolated testicular capsule of the rat (Sanchez, Manso et al. 1989).

Calcium Channels

The calcium-release-activated Ca^{2+}-channel, CRAC or ICRAC (Vig and Kinet 2007), is a highly Ca^{2+}-selective ion channel, activated on depletion of either intracellular Ca^{2+} levels or intracellular Ca^{2+} stores. The protein was identified by Yue and coworkers (Yue, Peng et al. 2001) and one of its characteristic properties was that is was blocked by lanthanum (Aussel, Marhaba et al. 1996; Vazquez, de Boland et al. 1997; Fasolato and Nilius 1998; Chang, Di Capite et al. 2008). Many cells have an electrogenic Ca^{2+} influx pathway based on so-called store-operated channels. The single transmembrane-spanning Ca^{2+}-binding protein, STIM1, has been proposed to function as a Ca^{2+} sensor that links the endoplasmic reticulum to the activation of these channels; these channels can be blocked by lanthanum (Ma, Smith et al. 2000; Koh, Lee et al. 2009). Lanthanum also interacts with the L-type voltage-gated Ca^{2+} channel in pancreatic β-cells; lanthanum does not enter the cells but still supports glucose-induced insulin secretion, an effect similar to that of Ca^{2+} (Trus, Corkey et al. 2007).

Effects of Lanthanum on Other Ion Channels

Lanthanum also blocks K^+ channels, e.g., in *Xenopus* A6 kidney epithelial cells (De Smet, Li et al. 1998), bovine adrenal zona fasciculata (Enyeart, Xu et al. 2002) and in *Drosophila* neurons (Alshuaib and Mathew 2004),Cl^- channels in *Xenopus* oocytes (Tokimasa and North 1996; Lomax, Herrero et al. 1998), and Na^+ currents in canine cardiac Purkinje cells (Sheets and Hanck 1992).

Chapter VI

Uptake of Lanthanum Salts in the Body

When lanthanum is administered to experimental animals, it ends up mainly in hepatocytes and enterocytes; in these cells, it has been shown by a variety of microanalytical techniques (e.g., X-ray microanalysis and secondary ion mass spectroscopy) that lanthanum is sequestered in the lysosomes, generally together with phosphorus, i.e., as phosphates (Berry 1996; Manoubi-Tekaya, Houcine et al. 2000; Floren, Tekaya et al. 2001; Fehri, Ayadi et al. 2005; Tekaya, Ayadi et al. 2005; Yang, Schryvers et al. 2006). Since superficial enterocytes have a life-time of only a few days, the lysosomal sequestration provides the body with protection against the possibly harmful effects of lanthanum. When, on the other hand, lanthanum is installed in the trachea, the metal remains mostly in the lung, and localizes in macrophages as high electron-dense granular inclusions in lysosomes and on the cell surface and basement membranes of type I pneumocytes among lung cells (Suzuki, Kobayashi et al. 1992).

Chapter VII

Lanthanum Carbonate in the Treatment of Hyperphosphatemia in End Stage Renal Disease

Hyperphosphatemia is common in patients with end-stage renal disease (ESRD), since a large fraction (60-70%) of dietary phosphorus is absorbed and normally excreted by the kidneys, and as kidney function deteriorates, less phosphorus is excreted by the kidneys (Emmett 2004). Dietary restrictions have insufficient effect. The condition may have serious consequences. Hyperphosphatemia stimulates parathyroid hormone and suppresses vitamin D3 production, and thus induces hyperparathyroid bone disease. In addition, it leads to myocardial and vascular calcification and cardiac microcirculatory abnormalities, which results in cardiac causes of death. Phosphate levels hence should be controlled early in the course of chronic renal failure in order to avoid secondary hyperparathyroidism, and cardiovascular and soft tissue calcifications. Because phosphate is not easily removed by dialysis, traditionally, phosphate level control has been done by oral phosphate binders, but up to a few years ago, none of the phosphate binders used clinically was entirely satisfactory (Albaaj and Hutchison 2003) . Aluminum-based agents are associated with bone toxicity, renal osteodystrophy and encephalopathy, and calcium-based agents increase the risk of hypercalcaemia and cardiovascular calcification (Spasovski, Massy et al. 2009). Sevelamer hydrochloride (trade names Renagel or Renvela) is a partial hydrochloride salt being present as approximately 40% amine hydrochloride and 60% sevelamer base (a copolymer of 2-(chloromethyl)oxirane and prop-2-en-1-amine). The amine groups of sevelamer are partially protonated in the intestine and bind phosphate. Sevelamer represented a step forward in the management of hyperphosphatemia, as it had a decreased risk of vascular calcification (Bleyer 2003) but has several drawbacks, in addition to high cost and a large pill burden (Albaaj and Hutchison 2003; Mohammed and Hutchison 2008). Adverse effects of sevalamer include hypotension, hypertension, nausea and vomiting, dyspepsia, diarrhea, flatulence, constipation, and metabolic acidosis; these effects often result in discontinuation of treatment (Shigematsu 2008). The most recent non-calcium, non-aluminum phosphate binder is lanthanum carbonate (trade name Fosrenol) (D'Haese, Spasovski et al. 2003).

The first stage III studies (D'Haese, Spasovski et al. 2003) showed that lanthanum carbonate (LC) was a poorly absorbed, well-tolerated, and efficient phosphate binder. The absolute bioavailability of lanthanum, administered as lanthanum carbonate, was found to be extremely low in a later study (Pennick, Dennis et al. 2006). LC-treated dialysis patients showed almost no evolution toward low bone turnover over one year, unlike calcium carbonate (CC)-treated patients, nor did they experience any aluminum-like effects of the drug on bone, which may be due to the low rates of intestinal absorption of lanthanum. In an animal study, DeBroe and coworkers (De Broe and D'Haese 2004) showed that no adverse effects on bone were seen in healthy animals receiving lanthanum carbonate. Although very high doses of lanthanum (1000-2000 mg/kg) affected bone mineralization, this was not due to a direct toxic effect on bone, but secondary to phosphate depletion induced by lanthanum (Behets, Verberckmoes et al. 2004). As with any gastro-intestinal phosphate-binding agent, this effect could be reversed with a phosphate-supplemented diet. The same group also showed that patients treated with lanthanum carbonate for 1 year did not experience any of the aluminum-like toxic effects on bone expressed as either osteomalacia or adynamic bone disease (Behets, Dams et al. 2004; De Broe and D'Haese 2004), and Freemont and Malluche (Freemont and Malluche 2005) confirmed that, with regard to effects on bone, LC was superior to CC. The absence of negative effects of LC on bone was again confirmed in a study on ESRD patients (Freemont, Hoyland et al. 2005). The initial studies in Europe (Albaaj and Hutchison 2003; Behets, Dams et al. 2004; Hutchison, Speake et al. 2004), the U.S.A. (Finn, Joy et al. 2004), China (Chiang, Chen et al. 2005) and Japan (Shigematsu 2008) confirmed that LC was an effective and well-tolerated agent for the short-term treatment of hyperphosphatemia in patients with ESRD. Subsequent studies confirmed these positive findings for longer-term studies of 3 years (Hutchison, Maes et al. 2006) and 6 years (Hutchison, Barnett et al. 2008; Hutchison, Barnett et al. 2009), stressed the cost-effectiveness of lanthanum carbonate treatment for hyperphosphatemia in ESRD patients (Brennan, Akehurst et al. 2007) and confirmed the superiority of LC over alternative drugs (Sprague 2007). One of the advantages of lanthanum carbonate is the low tablet burden (Hutchison and Laville 2008; Mehrotra, Martin et al. 2008). Another advantage of lanthanum carbonate over calcium-based drugs is that rather than increasing the risk of hypercalcaemia and cardiovascular calcification, lanthanum carbonate actually prevents arterial calcification (in animal experiments) (Neven, Dams et al. 2009). Nevertheless, a recent meta-analysis of published clinical trials (Navaneethan, Palmer et al. 2009) concluded that there are as yet insufficient data to establish the comparative superiority of non-calcium-binding agents, such as lanthanum carbonate, over calcium-containing phosphate binders for such important patient-level outcomes as all-cause mortality and cardiovascular end points.

Lanthanum carbonate has effects on some endocrine glands. Like sevelamer hydrochloride and calcium carbonate, lanthanum carbonate was shown to inhibit intestinal uptake of thyroid hormone (levothyroxine), which should be taken into account in patients taking both medications (Weitzman, Ginsburg et al. 2009). In rats with adenine-induced chronic kidney disease, where parathyroid hormone (PTH) mRNA is more stable, the binding of K-homology splicing regulatory protein (KSRP), a PTH mRNA destabilizing protein, is increased by treatment with lanthanum, correlating with decreased PTH gene expression. Hence, treatment of patients with lanthanum carbonate lowers PTH mRNA stability through KSRP-mediated recruitment of a degradation complex to the PTH mRNA, thereby decreasing PTH expression (Nechama, Ben-Dov et al. 2009).

The liver plays an important role in the metabolism of lanthanum carbonate. Most of the absorbed lanthanum is excreted in the bile (Damment and Pennick 2007). Bervoets and coworkers (Bervoets, Behets et al. 2009) found that in the liver lanthanum was located in lysosomes and in the biliary canal but not in any other cellular organelles, which suggests that lanthanum is transported and eliminated by the liver via a transcellular, endosomal-lysosomal-biliary canicular transport route. Feeding rats with chronic renal failure orally with lanthanum resulted in a doubling of the liver levels compared to rats with normal renal function, but the serum levels were similar in both animal groups. These levels plateaued after 6 weeks at a concentration below 3 µg/g in both groups. When lanthanum was administered intravenously, thereby bypassing the gastrointestinal tract-portal vein pathway, no difference in liver levels was found between rats with and without renal failure. This can be explained by an increased gastrointestinal permeability or absorption of oral lanthanum in uremia. Whether the findings of Bervoets et al. (2009) have any practical consequences for the treatment of dialysis patients is still a matter of discussion (Hutchison 2009)..

Outside the gut, the highest lanthanum concentrations are found in the mesenteric lymph nodes. Davis and Abraham (Davis and Abraham 2009) observed lanthanum deposition in a mesenteric lymph node at autopsy of a 38-year-old female ESRD patient 3 years following lanthanum carbonate administration.

The safety of lanthanum carbonate has been debated with great intensity. In particular, it was feared that lanthanum could have the same effect as aluminum, namely the death of many uremic patients due to toxic effects on the central nervous system (hallucinations, seizures, dementia) and bone (osteomalacia, osteodystrophy, and bone pain) (Canavese, Mereu et al. 2005; 2007). It appears that negative results originate (mainly) from animal (rodent) studies. Lacour and coworkers (Lacour, Lucas et al. 2005) and Slatopolsky and corworkers (Slatopolsky, Liapis et al. 2005) concluded that oral administration of LC to normal rats led to a more than 10-fold increase of tissue lanthanum content in at least some organs (liver, lung, and kidney) and that this increase is further enhanced by the uremic state. Ben-Dov coworkers (Ben-Dov, Pappo et al. 2007) on the other hand, found no liver toxicity neither in uremic rats nor in normal rats. In other rat experiments, it was found that lanthanum levels in the brain and heart fluctuated near its detection limit with long-term treatment (20 weeks) having no effect on organ weight, liver enzyme activities, or liver histology. Damment and coworkers (Damment and Shen 2005; Bervoets, Behets et al. 2009) confirmed in a study on rats the findings of Behets and coworkers (Behets, Verberckmoes et al. 2004), that the effect on lanthanum on bone was due to phosphate-depletion, not to bone toxicity. In another study of rats, it was shown that the lanthanum-induced mineralization defect in bone is normalized after arrest of lanthanum administration (Bervoets, Oste et al. 2006). Again in rats, it was observed that lanthanum exposure could significantly impair the behavioral performance, possibly due to significant changes in the distribution of brain elements such as Ca, Fe and Zn, and that, moreover, lanthanum significantly inhibited the activity of Ca^{2+}-ATPase; the function of the central cholinergic system was also noticeably disturbed and the contents of some monoamines neurotransmitters were significantly decreased (Feng, Xiao et al. 2006). Both *in vitro* (in cell cultures) and *in vivo* (in rat) tests showed that lanthanum was not genotoxic, and hence unlikely to present a problem in clinical use (Damment, Beevers et al. 2005).

Despite the mixed results in rats, clinical studies (on humans) confirmed that lanthanum carbonate treatment was safe (Finn 2006). Although cognitive function was found to decrease in hemodialysis patients during a 2-year study of lanthanum carbonate, this effect was not different from that observed in patients on standard therapy (Altmann, Barnett et al. 2007). On the other hand, Muller and coworkers (Muller, Chantrel et al. 2009) reported a case of confusion in a lanthanum carbonate-treated elderly women, in whom the confusion was resolved after suspension of the lanthanum treatment. A recent study (Damment, Cox et al. 2009) found that brain deposition of lanthanum is a contamination artifact caused by transfer of lanthanum from cranial skin to brain as animals are manipulated during autopsy, and recommended that dietary administration should be avoided in distribution studies of trace elements due to the high contamination risk.

In conclusion, although it may be too early for a final judgment, it appears that lanthanum carbonate may well be the treatment of choice for patients with hyperphosphatemia in end-stage renal disease.

Acknowledgments

The original research reviewed in this paper was financially supported by the Swedish Research Council and the Swedish Heart Lung Foundation.

References

(2007). "Lanthanum: new drug. Hyperphosphataemia in dialysis patients: more potential problems than benefits." *Prescrire Int.* 16(88): 47-50.

Agarwal, A. and A. P. Hoffer (1989). "Ultrastructural studies on the development of the blood-epididymis barrier in immature rats." *J. Androl.* 10(6): 425-31.

Albaaj, F. and A. Hutchison (2003). "Hyperphosphataemia in renal failure: causes, consequences and current management." *Drugs.* 63(6): 577-96.

Alshuaib, W. B. and M. V. Mathew (2004). "Blocking effect of lanthanum on delayed-rectifier K+ current in Drosophila neurons." *Int. J. Neurosci.* 114(5): 639-50.

Altmann, P., M. E. Barnett, et al. (2007). "Cognitive function in Stage 5 chronic kidney disease patients on hemodialysis: no adverse effects of lanthanum carbonate compared with standard phosphate-binder therapy." *Kidney Int.* 71(3): 252-9.

Alton, E. W., S. D. Manning, et al. (1991). "Characterization of a Ca(2+)-dependent anion channel from sheep tracheal epithelium incorporated into planar bilayers." *J. Physiol.* 443: 137-59.

Atagi, Y., H. Ikadai, et al. (1993). "Testicular disruption in the As (aspermia) mutant rat, with special reference to the aggregate of ribosomes." *J. Vet. Med. Sci.* 55(2): 301-6.

Aussel, C., R. Marhaba, et al. (1996). "Submicromolar La3+ concentrations block the calcium release-activated channel, and impair CD69 and CD25 expression in CD3- or thapsigargin-activated Jurkat cells." *Biochem. J.* 313 (Pt 3): 909-13.

Balda, M. S., C. Flores-Maldonado, et al. (2000). "Multiple domains of occludin are involved in the regulation of paracellular permeability." *J. Cell Biochem.* 78(1): 85-96.

Behets, G. J., G. Dams, et al. (2004). "Does the phosphate binder lanthanum carbonate affect bone in rats with chronic renal failure?" *J. Am. Soc. Nephrol.* 15(8): 2219-28.

Behets, G. J., S. C. Verberckmoes, et al. (2004). "Lanthanum carbonate: a new phosphate binder." *Curr. Opin. Nephrol. Hypertens.* 13(4): 403-9.

Ben-Dov, I. Z., O. Pappo, et al. (2007). "Lanthanum carbonate decreases PTH gene expression with no hepatotoxicity in uraemic rats." *Nephrol. Dial. Transplant.* 22(2): 362-8.

Bergmann, M., D. Nashan, et al. (1989). "Pattern of compartmentation in human seminiferous tubules showing dislocation of spermatogonia." *Cell Tissue Res.* 256(1): 183-90.

Bergmann, M. and J. Schindelmeiser (1987). "Development of the blood-testis barrier in the domestic fowl (Gallus domesticus)." *Int. J. Androl.* 10(2): 481-8.

Berry, J. P. (1996). "The role of lysosomes in the selective concentration of mineral elements. A microanalytical study." *Cell Mol. Biol.* (Noisy-le-grand) 42(3): 395-411.

Bervoets, A. R., G. J. Behets, et al. (2009). "Hepatocellular transport and gastrointestinal absorption of lanthanum in chronic renal failure." *Kidney Int.* 75(4): 389-98.

Bervoets, A. R., L. Oste, et al. (2006). "Development and reversibility of impaired mineralization associated with lanthanum carbonate treatment in chronic renal failure rats." *Bone.* 38(6): 803-10.

Bishop, M. A. (1985). "Evidence for tight junctions between odontoblasts in the rat incisor." *Cell Tissue Res.* 239(1): 137-40.

Bjarnason, I., A. MacPherson, et al. (1995). "Intestinal permeability: an overview." *Gastroenterology.* 108(5): 1566-81.

Blaug, S., K. Hybiske, et al. (2001). "ENaC- and CFTR-dependent ion and fluid transport in mammary epithelia." *Am. J. Physiol. Cell Physiol.* 281(2): C633-48.

Bleyer, A. J. (2003). "Phosphate binder usage in kidney failure patients." *Expert Opin. Pharmacother.* 4(6): 941-7.

Braga, J., J. M. Desterro, et al. (2004). "Intracellular macromolecular mobility measured by fluorescence recovery after photobleaching with confocal laser scanning microscopes." *Mol. Biol. Cell.* 15(10): 4749-60.

Brennan, A., R. Akehurst, et al. (2007). "The cost-effectiveness of lanthanum carbonate in the treatment of hyperphosphatemia in patients with end-stage renal disease." *Value Health.* 10(1): 32-41.

Brommundt, G. and F. Kavaler (1987). "La3+, Mn2+, and Ni2+ effects on Ca2+ pump and on Na+-Ca2+ exchange in bullfrog ventricle." *Am. J. Physiol.* 253(1 Pt 1): C45-51.

Bustamante, J. C. and B. P. Setchell (2000). "The permeability of the microvasculature of the perfused rat testis to small hydrophilic substances." *J. Androl.* 21(3): 444-51.

Caldwell, R. B. and B. J. McLaughlin (1983). "Permeability of retinal pigment epithelial cell junctions in the dystrophic rat retina." *Exp. Eye Res.* 36(3): 415-27.

Caldwell, R. B., S. M. Slapnick, et al. (1985). "Lanthanum and freeze-fracture studies of retinal pigment epithelial cell junctions in the streptozotocin diabetic rat." *Curr. Eye Res.* 4(3): 215-27.

Cambrosio Mann, M., A. E. Friess, et al. (2003). "Blood-tissue barriers in the male reproductive tract of the dog: a morphological study using lanthanum nitrate as an electron-opaque tracer." *Cells Tissues Organs.* 174(4): 162-9.

Canavese, C., C. Mereu, et al. (2005). "Blast from the past: the aluminum's ghost on the lanthanum salts." *Curr. Med. Chem.* 12(14): 1631-6.

Cavicchia, J. C. and F. L. Sacerdote (1988). "Topography of the rat blood-testis barrier after intratubular administration of intercellular tracers." *Tissue Cell.* 20(4): 577-86.

Cavicchia, J. C. and F. L. Sacerdote (1991). "Correlation between blood-testis barrier development and onset of the first spermatogenic wave in normal and in busulfan-treated rats: a lanthanum and freeze-fracture study." *Anat. Rec.* 230(3): 361-8.

Cereijido, M., R. G. Contreras, et al. (2007). "New diseases derived or associated with the tight junction." *Arch. Med. Res.* 38(5): 465-78.

Cetinel, S., B. Cetinel, et al. (2003). "Indomethacin-induced morphologic changes in the rat urinary bladder epithelium." *Urology.* 61(1): 236-42.

Chang, W. C., J. Di Capite, et al. (2008). "Local Ca2+ influx through Ca2+ release-activated Ca2+ (CRAC) channels stimulates production of an intracellular messenger and an intercellular pro-inflammatory signal." *J. Biol. Chem.* 283(8): 4622-31.

Che, Y., Y. Cui, et al. (2009). "Effects of lanthanum chloride administration in prenatal stage on one-trial passive avoidance learning in chicks." *Biol. Trace Elem. Res.* 127(1): 37-44.

Chiang, S. S., J. B. Chen, et al. (2005). "Lanthanum carbonate (Fosrenol) efficacy and tolerability in the treatment of hyperphosphatemic patients with end-stage renal disease." *Clin. Nephrol.* 63(6): 461-70.

Claude, P. and D. A. Goodenough (1973). "Fracture faces of zonulae occludentes from "tight" and "leaky" epithelia." *J. Cell Biol.* 58(2): 390-400.

Colegio, O. R., C. Van Itallie, et al. (2003). "Claudin extracellular domains determine paracellular charge selectivity and resistance but not tight junction fibril architecture." *Am. J. Physiol. Cell Physiol.* 284(6): C1346-54.

Curtis, M. J., D. M. Quastel, et al. (1986). "Lanthanum as a surrogate for calcium in transmitter release at mouse motor nerve terminals." *J. Physiol.* 373: 243-60.

D'Haese, P. C., G. B. Spasovski, et al. (2003). "A multicenter study on the effects of lanthanum carbonate (Fosrenol) and calcium carbonate on renal bone disease in dialysis patients." *Kidney Int. Suppl.* (85): S73-8.

Damment, S. J., C. Beevers, et al. (2005). "Evaluation of the potential genotoxicity of the phosphate binder lanthanum carbonate." *Mutagenesis.* 20(1): 29-37.

Damment, S. J., A. G. Cox, et al. (2009). "Dietary administration in rodent studies distorts the tissue deposition profile of lanthanum carbonate; brain deposition is a contamination artefact?" *Toxicol. Lett.* 188(3): 223-9.

Damment, S. J. and M. Pennick (2007). "Systemic lanthanum is excreted in the bile of rats." *Toxicol. Lett.* 171(1-2): 69-77.

Damment, S. J. and V. Shen (2005). "Assessment of effects of lanthanum carbonate with and without phosphate supplementation on bone mineralization in uremic rats." *Clin. Nephrol.* 63(2): 127-37.

Davis, R. L. and J. L. Abraham (2009). "Lanthanum deposition in a dialysis patient." *Nephrol. Dial. Transplant.* 24(10): 3247-50.

Daviskas, E., S. D. Anderson, et al. (1997). "Inhalation of dry-powder mannitol increases mucociliary clearance." *Eur. Respir. J.* 10(11): 2449-54.

Daviskas, E., S. D. Anderson, et al. (1996). "Inhalation of hypertonic saline aerosol enhances mucociliary clearance in asthmatic and healthy subjects." *Eur. Respir. J.* 9(4): 725-32.

Daviskas, E., M. Robinson, et al. (2002). "Osmotic stimuli increase clearance of mucus in patients with mucociliary dysfunction." *J. Aerosol. Med.* 15(3): 331-41.

De Broe, M. E. and P. C. D'Haese (2004). "Improving outcomes in hyperphosphataemia." *Nephrol. Dial. Transplant.* 19 Suppl 1: i14-8.

de Jong, E. W., L. Bosch, et al. (1976). "Isolation and characterization of a Ca2+ -binding polysaccharide associated with coccoliths of Emiliania huxleyi (Lohmann) Kamptner." *Eur. J. Biochem.* 70(2): 611-21.

De Smet, P., J. Li, et al. (1998). "Hypotonicity activates a lanthanide-sensitive pathway for K+ release in A6 epithelia." *Am. J. Physiol.* 275(1 Pt 1): C189-99.

Demerjian, M., D. A. Crumrine, et al. (2006). "Barrier dysfunction and pathogenesis of neutral lipid storage disease with ichthyosis (Chanarin-Dorfman syndrome)." *J. Invest. Dermatol.* 126(9): 2032-8.

Dimitratos, S. D., D. F. Woods, et al. (1999). "Signaling pathways are focused at specialized regions of the plasma membrane by scaffolding proteins of the MAGUK family." *Bioessays.* 21(11): 912-21.

Doggenweiler, C. F. and S. Frenk (1965). "Staining Properties of Lanthanum on Cell Membranes." *Proc. Natl. Acad. Sci. U. S. A.* 53: 425-30.

Dogic, Z., K. R. Purdy, et al. (2004). "Isotropic-nematic phase transition in suspensions of filamentous virus and the neutral polymer Dextran." *Phys. Rev. E Stat. Nonlin. Soft Matter Phys.* 69(5 Pt 1): 051702.

Donaldson, S. H., W. D. Bennett, et al. (2006). "Mucus clearance and lung function in cystic fibrosis with hypertonic saline." *N. Engl. J. Med.* 354(3): 241-50.

Drubin, D. G. and W. J. Nelson (1996). "Origins of cell polarity." *Cell.* 84(3): 335-44.

Ebnet, K., A. Suzuki, et al. (2001). "The cell polarity protein ASIP/PAR-3 directly associates with junctional adhesion molecule (JAM)." *Embo J.* 20(14): 3738-48.

Elkins, M. R., M. Robinson, et al. (2006). "A controlled trial of long-term inhaled hypertonic saline in patients with cystic fibrosis." *N. Engl. J. Med.* 354(3): 229-40.

Emmett, M. (2004). "A comparison of clinically useful phosphorus binders for patients with chronic kidney failure." *Kidney Int. Suppl.* (90): S25-32.

Enyeart, J. J., L. Xu, et al. (2002). "Dual actions of lanthanides on ACTH-inhibited leak K(+) channels." *Am. J. Physiol. Endocrinol. Metab.* 282(6): E1255-66.

Erlij, D. and A. Martinez-Palomo (1972). "Opening of tight junctions in frog skin by hypertonic urea solutions." *J. Membr. Biol.* 9(3): 229-40.

Fasolato, C. and B. Nilius (1998). "Store depletion triggers the calcium release-activated calcium current (ICRAC) in macrovascular endothelial cells: a comparison with Jurkat and embryonic kidney cell lines." *Pflugers Arch.* 436(1): 69-74.

Fehri, E., A. Ayadi, et al. (2005). "[Lanthanides and microanalysis. Effects of oral administration of two lanthanides: ultrastructural and microanalytical study]." *Arch. Inst. Pasteur Tunis.* 82(1-4): 59-67.

Feng, L., H. Xiao, et al. (2006). "Neurotoxicological consequence of long-term exposure to lanthanum." *Toxicol. Lett.* 165(2): 112-20.

Finn, W. F. (2006). "Lanthanum carbonate versus standard therapy for the treatment of hyperphosphatemia: safety and efficacy in chronic maintenance hemodialysis patients." *Clin. Nephrol.* 65(3): 191-202.

Finn, W. F., M. S. Joy, et al. (2004). "Efficacy and safety of lanthanum carbonate for reduction of serum phosphorus in patients with chronic renal failure receiving hemodialysis." *Clin. Nephrol.* 62(3): 193-201.

Fitzpatrick, L. A. (1990). "Differences in the actions of calcium versus lanthanum to influence parathyroid hormone release." *Endocrinology.* 127(2): 711-5.

Floren, C., L. Tekaya, et al. (2001). "Analytical microscopy observations of rat enterocytes after oral administration of soluble salts of lanthanides, actinides and elements of group III-A of the periodic chart." *Cell Mol. Biol.* (Noisy-le-grand) 47(3): 419-25.

Freemont, A. J., J. A. Hoyland, et al. (2005). "The effects of lanthanum carbonate and calcium carbonate on bone abnormalities in patients with end-stage renal disease." *Clin. Nephrol.* 64(6): 428-37.

Freemont, T. and H. H. Malluche (2005). "Utilization of bone histomorphometry in renal osteodystrophy: demonstration of a new approach using data from a prospective study of lanthanum carbonate." *Clin. Nephrol.* 63(2): 138-45.

Fries, W., E. Mazzon, et al. (1999). "Experimental colitis increases small intestine permeability in the rat." *Lab. Invest.* 79(1): 49-57.

Furuse, M., M. Itoh, et al. (1994). "Direct association of occludin with ZO-1 and its possible involvement in the localization of occludin at tight junctions." *J. Cell Biol.* 127(6 Pt 1): 1617-26.

Ganong, W. F. (2000). "Circumventricular organs: definition and role in the regulation of endocrine and autonomic function." *Clin. Exp. Pharmacol. Physiol.* 27(5-6): 422-7.

Gotow, T. and P. H. Hashimoto (1979). "Fine structure of the ependyma and intercellular junctions in the area postrema of the rat." *Cell Tissue Res.* 201(2): 207-25.

Gould, R. J., K. M. Murphy, et al. (1982). "[3H]nitrendipine-labeled calcium channels discriminate inorganic calcium agonists and antagonists." *Proc. Natl. Acad. Sci. U. S. A.* 79(11): 3656-60.

Gruenert, D. C., M. Willems, et al. (2004). "Established cell lines used in cystic fibrosis research." *J. Cyst. Fibros. 3 Suppl.* 2: 191-6.

Guan, X., T. Inai, et al. (2005). "Segment-specific expression of tight junction proteins, claudin-2 and -10, in the rat epididymal epithelium." *Arch. Histol. Cytol.* 68(3): 213-25.

Gumbiner, B. M. (1993). "Breaking through the tight junction barrier." *J. Cell Biol.* 123(6 Pt 2): 1631-3.

Gylfe, E., R. Larsson, et al. (1986). "Calcium-activated calcium permeability in parathyroid cells." *FEBS Lett.* 205(1): 132-6.

Hatier, R. and G. Grignon (1986). "Ultrastructural study of the Sertoli cell and the limiting membrane in the seminiferous tubule of the adult cryptorchid rat." *Virchows Arch. B Cell Pathol. Incl. Mol. Pathol.* 52(4): 305-18.

Hawkins, C. P., P. M. Munro, et al. (1992). "Metabolically dependent blood-brain barrier breakdown in chronic relapsing experimental allergic encephalomyelitis." *Acta Neuropathol.* 83(6): 630-5.

Hay, D. W. and R. M. Wadsworth (1982). "Effects of some organic calcium antagonists and other procedures affecting Ca2+ Translocation on KCl-induced contractions in the rat vas deferens." *Br. J. Pharmacol.* 76(1): 103-13.

Hirsh, A. J. (2002). "Altering airway surface liquid volume: inhalation therapy with amiloride and hyperosmotic agents." *Adv. Drug Deliv. Rev.* 54(11): 1445-62.

Hochman, J. H., J. A. Fix, et al. (1994). "In vitro and in vivo analysis of the mechanism of absorption enhancement by palmitoylcarnitine." *J. Pharmacol. Exp. Ther.* 269(2): 813-22.

Holland, V. F., G. A. Zampighi, et al. (1989). "Morphology of fungiform papillae in canine lingual epithelium: location of intercellular junctions in the epithelium." *J. Comp. Neurol.* 279(1): 13-27.

Hua, X., L. P. Zhu, et al. (2009). "Effects of diagnostic contrast-enhanced ultrasound on permeability of placental barrier: a primary study." *Placenta.* 30(9): 780-4.

Huang, H. F., C. S. Yang, et al. (1988). "Disruption of sustentacular (Sertoli) cell tight junctions and regression of spermatogenesis in vitamin-A-deficient rats." *Acta Anat.* (Basel) 133(1): 10-5.

Huber, D., M. S. Balda, et al. (2000). "Occludin modulates transepithelial migration of neutrophils." *J. Biol. Chem.* 275(8): 5773-8.

Hull, J., W. Skinner, et al. (1998). "Elemental content of airway surface liquid from infants with cystic fibrosis." *Am. J. Respir. Crit. Care Med.* 157(1): 10-4.

Hutchison, A. J. (2009). "Lanthanum and phosphate: science, policy, and survival." *Kidney Int.* 75(4): 355-7.

Hutchison, A. J., M. E. Barnett, et al. (2008). "Long-term efficacy and safety profile of lanthanum carbonate: results for up to 6 years of treatment." *Nephron. Clin. Pract.* 110(1): c15-23.

Hutchison, A. J., M. E. Barnett, et al. (2009). "Lanthanum carbonate treatment, for up to 6 years, is not associated with adverse effects on the liver in patients with chronic kidney disease Stage 5 receiving hemodialysis." *Clin. Nephrol.* 71(3): 286-95.

Hutchison, A. J. and M. Laville (2008). "Switching to lanthanum carbonate monotherapy provides effective phosphate control with a low tablet burden." *Nephrol. Dial. Transplant.* 23(11): 3677-84.

Hutchison, A. J., B. Maes, et al. (2006). "Long-term efficacy and tolerability of lanthanum carbonate: results from a 3-year study." *Nephron. Clin. Pract.* 102(2): c61-71.

Hutchison, A. J., M. Speake, et al. (2004). "Reducing high phosphate levels in patients with chronic renal failure undergoing dialysis: a 4-week, dose-finding, open-label study with lanthanum carbonate." *Nephrol. Dial. Transplant.* 19(7): 1902-6.

Högman, M., A. C. Mork, et al. (2002). "Hypertonic saline increases tight junction permeability in airway epithelium." *Eur. Respir. J.* 20(6): 1444-8.

Ikeda, S., T. Mitaka, et al. (2003). "Tumor necrosis factor-alpha and interleukin-6 reduce bile canalicular contractions of rat hepatocytes." *Surgery.* 133(1): 101-9.

Itoh, M., M. Furuse, et al. (1999). "Direct binding of three tight junction-associated MAGUKs, ZO-1, ZO-2, and ZO-3, with the COOH termini of claudins." *J. Cell Biol.* 147(6): 1351-63.

Itoh, M., H. Sasaki, et al. (2001). "Junctional adhesion molecule (JAM) binds to PAR-3: a possible mechanism for the recruitment of PAR-3 to tight junctions." *J. Cell Biol.* 154(3): 491-7.

Ivanov, A. I., I. C. McCall, et al. (2006). "Microtubules regulate disassembly of epithelial apical junctions." *BMC Cell Biol.* 7: 12.

Jezernik, K. (1996). "Desquamation of urinary bladder epithelial cells." *Pflugers Arch.* 431(6 Suppl 2): R249-50.

Jiang, S. J., A. W. Chu, et al. (2007). "Ultraviolet B-induced alterations of the skin barrier and epidermal calcium gradient." *Exp. Dermatol.* 16(12): 985-92.

Kaur, C., W. S. Foulds, et al. (2008). "Blood-retinal barrier in hypoxic ischaemic conditions: basic concepts, clinical features and management." *Prog. Retin Eye Res.* 27(6): 622-47.

Kerper, L. E. and P. M. Hinkle (1997). "Cellular uptake of lead is activated by depletion of intracellular calcium stores." *J. Biol. Chem.* 272(13): 8346-52.

Kertschanska, S., B. Stulcova, et al. (2000). "Distensible transtrophoblastic channels in the rat placenta." *Placenta.* 21(7): 670-7.

Kiessling, F., J. Kartenbeck, et al. (1999). "Cell-cell contacts in the human cell line ECV304 exhibit both endothelial and epithelial characteristics." *Cell Tissue Res.* 297(1): 131-40.

Koh, S., K. Lee, et al. (2009). "STIM1 regulates store-operated Ca2+ entry in oocytes." *Dev. Biol.* 330(2): 368-76.

Korosec, P. and K. Jezernik (2000). "Early cellular and ultrastructural response of the mouse urinary bladder urothelium to ischemia." *Virchows Arch.* 436(4): 377-83.

Kozlova, I., V. Vanthanouvong, et al. (2005). "Elemental composition of airway surface liquid in the pig determined by x-ray microanalysis." *Am. J. Respir. Cell Mol. Biol.* 32(1): 59-64.

Kozlova, I., V. Vanthanouvong, et al. (2006). "Composition of airway surface liquid determined by X-ray microanalysis." *Ups. J. Med. Sci.* 111(1): 137-53.

Kreft, M. E., R. Romih, et al. (2002). "Antigenic and ultrastructural markers associated with urothelial cytodifferentiation in primary explant outgrowths of mouse bladder." *Cell. Biol. Int.* 26(1): 63-74.

Lacaz-Vieira, F. (1997). "Calcium site specificity. Early Ca2+-related tight junction events." *J. Gen. Physiol.* 110(6): 727-40.

Lacour, B., A. Lucas, et al. (2005). "Chronic renal failure is associated with increased tissue deposition of lanthanum after 28-day oral administration." *Kidney Int.* 67(3): 1062-9.

Lane, N. J., J. B. Harrison, et al. (1981). "A vertebrate-like blood--brain barrier, with intraganglionic blood channels and occluding junctions, in the scorpion." *Tissue Cell.* 13(3): 557-76.

Leslie, R. A. (1975). "The effects of ionic lanthanum and hypertonic physiological salines on the nervous systems of larval and adult stick insects." *J. Cell Sci.* 18(2): 271-86.

Lettvin, J. Y., W. F. Pickard, et al. (1964). "A Theory of Passive Ion Flux through Axon Membranes." *Nature.* 202: 1338-9.

Levy, S., V. Serre, et al. (1999). "The effects of aging on the seminiferous epithelium and the blood-testis barrier of the Brown Norway rat." *J. Androl.* 20(3): 356-65.

Lo, W. K. (1987). "In vivo and in vitro observations on permeability and diffusion pathways of tracers in rat and frog lenses." *Exp. Eye Res.* 45(3): 393-406.

Lomax, R. B., C. J. Herrero, et al. (1998). "Capacitative Ca2+ entry into Xenopus oocytes is sensitive to omega-conotoxins GVIA, MVIIA and MVIIC." *Cell Calcium.* 23(4): 229-39.

Lopez, M. L., P. Fuentes, et al. (1997). "Regional differentiation of the blood-epididymis barrier in stallion (Equus caballus)." *J. Submicrosc. Cytol. Pathol.* 29(3): 353-63.

Lora, L., E. Mazzon, et al. (1997). "Hepatocyte tight-junctional permeability is increased in rat experimental colitis." *Gastroenterology.* 113(4): 1347-54.

Ma, R., S. Smith, et al. (2000). "Store-operated Ca(2+) channels in human glomerular mesangial cells." *Am. J. Physiol. Renal Physiol.* 278(6): F954-61.

Machen, T. E., D. Erlij, et al. (1972). "Permeable junctional complexes. The movement of lanthanum across rabbit gallbladder and intestine." *J. Cell Biol.* 54(2): 302-12.

MacKenzie, M. L., M. N. Ghabriel, et al. (1987). "The blood-nerve barrier: an in vivo lanthanum tracer study." *J. Anat.* 154: 27-37.

Madara, J. L. and J. S. Trier (1982). "Structure and permeability of goblet cell tight junctions in rat small intestine." *J. Membr. Biol.* 66(2): 145-57.

Mandel, L. J., R. Bacallao, et al. (1993). "Uncoupling of the molecular 'fence' and paracellular 'gate' functions in epithelial tight junctions." *Nature.* 361(6412): 552-5.

Manoubi-Tekaya, L., N. Houcine, et al. (2000). "[Role of lysosomes in the intracellular mineral elements: the case of lanthane]." *Tunis Med.* 78(3): 195-200.

Martin, M. V., J. Appleton, et al. (1987). "The effect of Candida albicans on the permeability of rat palatal epithelium: an ultrastructural and biochemical study." *J. Med. Vet. Mycol.* 25(1): 19-28.

Mazzon, E. and S. Cuzzocrea (2007). "Absence of functional peroxisome proliferator-activated receptor-alpha enhanced ileum permeability during experimental colitis." *Shock.* 28(2): 192-201.

Mazzon, E., G. C. Sturniolo, et al. (2002). "Effect of stress on the paracellular barrier in the rat ileum." *Gut.* 51(4): 507-13.

Mead, R. H. and W. T. Clusin (1985). "Paradoxical electromechanical effect of lanthanum ions in cardiac muscle cells." *Biophys. J.* 48(5): 695-700.

Mehrotra, R., K. J. Martin, et al. (2008). "Higher strength lanthanum carbonate provides serum phosphorus control with a low tablet burden and is preferred by patients and physicians: a multicenter study." *Clin. J. Am. Soc. Nephrol.* 3(5): 1437-45.

Meyer, J. M., P. Mezrahid, et al. (1996). "Sertoli cell barrier dysfunction and spermatogenetic cycle breakdown in the human testis: a lanthanum tracer investigation." *Int. J. Androl.* 19(3): 190-8.

Meyer, R. A., D. McGinley, et al. (1986). "Effects of aspirin on tight junction structure of the canine gastric mucosa." *Gastroenterology.* 91(2): 351-9.

Mitic, L. L., C. M. Van Itallie, et al. (2000). "Molecular physiology and pathophysiology of tight junctions I. Tight junction structure and function: lessons from mutant animals and proteins." *Am. J. Physiol. Gastrointest. Liver Physiol.* 279(2): G250-4.

Miyoshi, J. and Y. Takai (2008). "Structural and functional associations of apical junctions with cytoskeleton." *Biochim. Biophys. Acta.* 1778(3): 670-91.

Mohammed, I. A. and A. J. Hutchison (2008). "Phosphate binding therapy in dialysis patients: focus on lanthanum carbonate." *Ther. Clin. Risk Manag.* 4(5): 887-93.

Mollgard, K. and N. R. Saunders (1986). "The development of the human blood-brain and blood-CSF barriers." *Neuropathol. Appl. Neurobiol.* 12(4): 337-58.

Mora-Galindo, J. (1986). "Maturation of tight junctions in guinea-pig cecal epithelium." *Cell Tissue Res.* 246(1): 169-75.

Morales, A. and J. C. Cavicchia (2002). "Spermatogenesis and blood-testis barrier in rats after long-term Vitamin A deprivation." *Tissue Cell.* 34(5): 349-55.

Morales, A., F. Mohamed, et al. (2007). "Apoptosis and blood-testis barrier during the first spermatogenic wave in the pubertal rat." *Anat. Rec.* (Hoboken) 290(2): 206-14.

Muller, C., F. Chantrel, et al. (2009). "A confusional state associated with use of lanthanum carbonate in a dialysis patient: a case report." *Nephrol. Dial. Transplant.* 24(10): 3245-7.

Murthy, R. C., D. K. Saxena, et al. (1991). "Ultrastructural observations in testicular tissue of chromium-treated rats." *Reprod. Toxicol.* 5(5): 443-7.

Nag, S. (1991). "Effect of atrial natriuretic factor on permeability of the blood-cerebrospinal fluid barrier." *Acta Neuropathol.* 82(4): 274-9.

Nag, S., D. M. Robertson, et al. (1982). "Intracerebral arteriolar permeability to lanthanum." *Am. J. Pathol.* 107(3): 336-41.

Nagy, Z., G. Mathieson, et al. (1979). "Blood-brain barrier opening to horseradish peroxidase in acute arterial hypertension." *Acta Neuropathol.* 48(1): 45-53.

Nagy, Z., H. M. Pappius, et al. (1979). "Opening of tight junctions in cerebral endothelium. I. Effect of hyperosmolar mannitol infused through the internal carotid artery." *J. Comp. Neurol.* 185(3): 569-78.

Naik, U. P. and K. Eckfeld (2003). "Junctional adhesion molecule 1 (JAM-1)." *J. Biol. Regul. Homeost. Agents.* 17(4): 341-7.

Nakagawa, Y., J. Cervos-Navarro, et al. (1985). "Tracer study on a paracellular route in experimental hydrocephalus." *Acta Neuropathol.* 65(3-4): 247-54.

Nathan, R. D., K. Kanai, et al. (1988). "Selective block of calcium current by lanthanum in single bullfrog atrial cells." *J. Gen. Physiol.* 91(4): 549-72.

Navaneethan, S. D., S. C. Palmer, et al. (2009). "Benefits and harms of phosphate binders in CKD: a systematic review of randomized controlled trials." *Am. J. Kidney Dis.* 54(4): 619-37.

Nechama, M., I. Z. Ben-Dov, et al. (2009). "Regulation of PTH mRNA stability by the calcimimetic R568 and the phosphorus binder lanthanum carbonate in CKD." *Am. J. Physiol. Renal. Physiol.* 296(4): F795-800.

Neven, E., G. Dams, et al. (2009). "Adequate phosphate binding with lanthanum carbonate attenuates arterial calcification in chronic renal failure rats." *Nephrol. Dial Transplant.* 24(6): 1790-9.

Nilsson, H., A. Dragomir, et al. (2007). "Effects of hyperosmotic stress on cultured airway epithelial cells." *Cell Tissue Res.* 330(2): 257-69.

Nilsson, H., A. Dragomir, et al. (2006). "A modified technique for the impregnation of lanthanum tracer to study the integrity of tight junctions on cells grown on a permeable substrate." *Microsc. Res. Tech.* 69(10): 776-83.

Nilsson, H. E., A. Dragomir, et al. (2009). "CFTR and tight junctions in cultured bronchial epithelial cells." *Exp. Mol. Pathol.*

Nusrat, A., J. R. Turner, et al. (2000). "Molecular physiology and pathophysiology of tight junctions. IV. Regulation of tight junctions by extracellular stimuli: nutrients, cytokines, and immune cells." *Am. J. Physiol. Gastrointest Liver Physiol.* 279(5): G851-7.

Nylander, O., L. Pihl, et al. (2003). "Hypotonicity-induced increases in duodenal mucosal permeability facilitates adjustment of luminal osmolality." *Am. J. Physiol. Gastrointest. Liver Physiol.* 285(2): G360-70.

Ohtake, K., T. Maeno, et al. (2003). "Poly-L-arginine enhances paracellular permeability via serine/threonine phosphorylation of ZO-1 and tyrosine dephosphorylation of occludin in rabbit nasal epithelium." *Pharm. Res.* 20(11): 1838-45.

Okinami, S., M. Ohkuma, et al. (1978). "Disruption of blood-retinal barrier at the retinal pigment epithelium after systemic urea injection." *Acta Ophthalmol.* (Copenh) 56(1): 27-39.

Okinami, S., M. Ohkuma, et al. (1976). "Kuhnt intermediary tissue as a barrier between the optic nerve and retina." Albrecht. *Von Graefes Arch. Klin. Exp. Ophthalmol.* 201(1): 57-67.

Orlando, R. C., E. R. Lacy, et al. (1992). "Barriers to paracellular permeability in rabbit esophageal epithelium." *Gastroenterology.* 102(3): 910-23.

Pedersen, O. O. (1980). "Increased vascular permeability in the rabbit iris induced by prostaglandin E1. An electron microscopic study using lanthanum as a tracer in vivo." *Albrecht. Von Graefes Arch. Klin. Exp. Ophthalmol.* 212(3-4): 199-205.

Pelletier, R. M. (1986). "Cyclic formation and decay of the blood-testis barrier in the mink (Mustela vison), a seasonal breeder." *Am. J. Anat.* 175(1): 91-117.

Pennick, M., K. Dennis, et al. (2006). "Absolute bioavailability and disposition of lanthanum in healthy human subjects administered lanthanum carbonate." *J. Clin. Pharmacol.* 46(7): 738-46.

Powis, D. A., C. L. Clark, et al. (1994). "Lanthanum can be transported by the sodium-calcium exchange pathway and directly triggers catecholamine release from bovine chromaffin cells." *Cell Calcium.* 16(5): 377-90.

Provan, S. D. and M. D. Miyamoto (1992). "Subcellular mechanism and site of action of ionic lanthanum at the motor nerve terminal." *Neuroreport.* 3(1): 101-4.

Qu, Y. Z., M. Li, et al. (2009). "Astragaloside IV attenuates cerebral ischemia-reperfusion-induced increase in permeability of the blood-brain barrier in rats." *Eur. J. Pharmacol.* 606(1-3): 137-41.

Ranaldi, G., R. Consalvo, et al. (2003). "Permeability characteristics of parental and clonal human intestinal Caco-2 cell lines differentiated in serum-supplemented and serum-free media." *Toxicol. In Vitro.* 17(5-6): 761-7.

Rassat, J., H. Robenek, et al. (1981). "Ultrastructural alterations in the mouse liver following vincristine administration." *J. Submicrosc. Cytol.* 13(3): 321-35.

Rassat, J., H. Robenek, et al. (1982). "Alterations of tight and gap junctions in mouse hepatocytes following administration of colchicine." *Cell Tissue Res.* 223(1): 187-200.

Ravens, U., E. Steinmann, et al. (1987). "Effects of gallopamil, nifedipine, Ni2+, and La3+ in guinea pig atria after a sudden increase in extracellular Ca2+ concentration." *J. Cardiovasc. Pharmacol.* 10(4): 462-73.

Raviola, E. and M. J. Karnovsky (1972). "Evidence for a blood-thymus barrier using electron-opaque tracers." *J. Exp. Med.* 136(3): 466-98.

Reichling, D. B. and A. B. MacDermott (1991). "Lanthanum actions on excitatory amino acid-gated currents and voltage-gated calcium currents in rat dorsal horn neurons." *J. Physiol.* 441: 199-218.

Relova, A. J. and G. M. Roomans (2001). "Effect of luminal osmolarity on ion content of connective tissue in rat trachea after epithelial damage." *Eur. Respir. J.* 18(5): 810-6.

Revel, J. P. and M. J. Karnovsky (1967). "Hexagonal array of subunits in intercellular junctions of the mouse heart and liver." *J. Cell Biol.* 33(3): C7-C12.

Robenek, H., J. Rassat, et al. (1982). "Ultrastructural study of cholestasis induced by longterm treatment with estradiol valerate. I. Tight junctional analysis and tracer experiments." *Virchows Arch. B. Cell Pathol. Incl. Mol. Pathol.* 40(2): 201-15.

Rom-Glas, A., C. Sandler, et al. (1992). "The nss mutation or lanthanum inhibits light-induced Ca2+ influx into fly photoreceptors." *J. Gen. Physiol.* 100(5): 767-81.

Sanchez, M., G. Manso, et al. (1989). "Interactions between oxytocin- and calcium-modifying agents in the rat testicular capsule in vitro." *Eur. J. Pharmacol.* 168(2): 169-77.

Schatzki, P. F. and A. Newsome (1975). "Neutralized lanthanum solution: a largely noncolloidal ultrastructural tracer." *Stain. Technol.* 50(3): 171-8.

Seki, T., S. Harada, et al. (2008). "Evaluation of the establishment of a tight junction in Caco-2 cell monolayers using a pore permeation model involving two different sizes." *Biol. Pharm. Bull.* 31(1): 163-6.

Shaklai, M. and M. Tavassoli (1977). "A modified technique to obtain uniform precipitation of lanthanum tracer in the extracellular space." *J. Histochem. Cytochem.* 25(8): 1013-5.

Sheets, M. F. and D. A. Hanck (1992). "Mechanisms of extracellular divalent and trivalent cation block of the sodium current in canine cardiac Purkinje cells." *J. Physiol.* 454: 299-320.

Shigematsu, T. (2008). "Lanthanum carbonate effectively controls serum phosphate without affecting serum calcium levels in patients undergoing hemodialysis." *Ther. Apher. Dial.* 12(1): 55-61.

Siu, M. K. and C. Y. Cheng (2004). "Extracellular matrix: recent advances on its role in junction dynamics in the seminiferous epithelium during spermatogenesis." *Biol. Reprod.* 71(2): 375-91.

Slatopolsky, E., H. Liapis, et al. (2005). "Progressive accumulation of lanthanum in the liver of normal and uremic rats." *Kidney Int.* 68(6): 2809-13.

Spasovski, G., Z. Massy, et al. (2009). "Phosphate metabolism in chronic kidney disease: from pathophysiology to clinical management." *Semin. Dial.* 22(4): 357-62.

Sprague, S. M. (2007). "A comparative review of the efficacy and safety of established phosphate binders: calcium, sevelamer, and lanthanum carbonate." *Curr. Med. Res Opin.* 23(12): 3167-75.

Sturniolo, G. C., W. Fries, et al. (2002). "Effect of zinc supplementation on intestinal permeability in experimental colitis." *J. Lab. Clin. Med.* 139(5): 311-5.

Summers, M. J., S. F. Crowe, et al. (1996). "Administration of lanthanum chloride following a reminder induces a transient loss of memory retrieval in day-old chicks." *Brain Res. Cogn. Brain Res.* 4(2): 109-19.

Sun, E. L. and B. Gondos (1986). "Formation of the blood-testis barrier in the rabbit." *Cell Tissue Res.* 243(3): 575-8.

Sutiagin, P. V. and A. S. Pylaev (1983). "[Identification of rat sinus node pacemaker cells by intracellular injection of lanthanum ions]." *Biull. Eksp. Biol. Med.* 95(5): 93-5.

Suzuki, K. T., E. Kobayashi, et al. (1992). "Localization and health effects of lanthanum chloride instilled intratracheally into rats." *Toxicology.* 76(2): 141-52.

Tekaya, L., A. Ayadi, et al. (2005). "Selective mineral elements concentration of the intestinal mucosa role of the lysosomes of duodenal enterocytes in the handling of mineral elements after intragastric administration." *Cell Mol. Biol.* (Noisy-le-grand) 51 Suppl: OL819-27.

Tisher, C. C. and W. E. Yarger (1973). "Lanthanum permeability of the tight junction (zonula occludens) in the renal tubule of the rat." *Kidney Int.* 3(4): 238-50.

Todd, B. A., C. Inman, et al. (2000). "Ionic permeability of the frog sciatic nerve perineurium: parallel studies of potassium and lanthanum penetration using electrophysiological and electron microscopic techniques." *J. Neurocytol.* 29(8): 551-67.

Tokimasa, T. and R. A. North (1996). "Effects of barium, lanthanum and gadolinium on endogenous chloride and potassium currents in Xenopus oocytes." *J. Physiol.* 496 (Pt 3): 677-86.

Trus, M., R. F. Corkey, et al. (2007). "The L-type voltage-gated Ca2+ channel is the Ca2+ sensor protein of stimulus-secretion coupling in pancreatic beta cells." *Biochemistry.* 46(50): 14461-7.

Tsukita, S. and M. Furuse (1999). "Occludin and claudins in tight-junction strands: leading or supporting players?" *Trends Cell Biol.* 9(7): 268-73.

Tsukita, S., M. Furuse, et al. (1999). "Structural and signalling molecules come together at tight junctions." *Curr. Opin. Cell Biol.* 11(5): 628-33.

Walsh, S. V., A. M. Hopkins, et al. (2000). "Modulation of tight junction structure and function by cytokines." *Adv. Drug Deliv. Rev.* 41(3): 303-13.

Van Itallie, C. M., J. Holmes, et al. (2008). "The density of small tight junction pores varies among cell types and is increased by expression of claudin-2." *J. Cell Sci.* 121(Pt 3): 298-305.

Wang, X. W., G. R. Ding, et al. (2008). "Effect of electromagnetic pulse exposure on permeability of blood-testicle barrier in mice." *Biomed. Environ. Sci.* 21(3): 218-21.

Wark, P. A., V. McDonald, et al. (2005). "Nebulised hypertonic saline for cystic fibrosis." *Cochrane Database Syst. Rev.* (3): CD001506.

Watson, C. J., C. J. Hoare, et al. (2005). "Interferon-gamma selectively increases epithelial permeability to large molecules by activating different populations of paracellular pores." *J. Cell Sci.* 118(Pt 22): 5221-30.

Vazquez, G., A. R. de Boland, et al. (1997). "Stimulation of Ca2+ release-activated Ca2+ channels as a potential mechanism involved in non-genomic 1,25(OH)2-vitamin D3-induced Ca2+ entry in skeletal muscle cells." *Biochem. Biophys. Res. Commun.* 239(2): 562-5.

Weber, J. E., T. T. Turner, et al. (1988). "Effects of cytochalasin D on the integrity of the Sertoli cell (blood-testis) barrier." *Am. J. Anat.* 182(2): 130-47.

Weitzman, S. P., K. C. Ginsburg, et al. (2009). "Colesevelam hydrochloride and lanthanum carbonate interfere with the absorption of levothyroxine." *Thyroid.* 19(1): 77-9.

Wendt-Gallitelli, M. F. and G. Isenberg (1985). "Extra- and intracellular lanthanum: modified calcium distribution, inward currents and contractility in guinea pig ventricular preparations." *Pflugers Arch.* 405(4): 310-22.

Vig, M. and J. P. Kinet (2007). "The long and arduous road to CRAC." *Cell Calcium.* 42(2): 157-62.

Winterhager, E. and W. Kuhnel (1985). "Diffusion barriers in the vaginal epithelium during the estrous cycle in guinea pigs." *Cell Tissue Res.* 241(2): 325-31.

Winters, S. L. and D. B. Yeates (1997). "Interaction between ion transporters and the mucociliary transport system in dog and baboon." *J. Appl. Physiol.* 83(4): 1348-59.

Winters, S. L. and D. B. Yeates (1997). "Roles of hydration, sodium, and chloride in regulation of canine mucociliary transport system." *J. Appl. Physiol.* 83(4): 1360-9.

Wittchen, E. S., J. Haskins, et al. (1999). "Protein interactions at the tight junction. Actin has multiple binding partners, and ZO-1 forms independent complexes with ZO-2 and ZO-3." *J. Biol. Chem.* 274(49): 35179-85.

Wu, X., G. Yuan, et al. (1992). "Sodium-dependent nucleoside transport in choroid plexus from rabbit. Evidence for a single transporter for purine and pyrimidine nucleosides." *J. Biol. Chem.* 267(13): 8813-8.

Xu, J. and E. A. Ling (1994). "Studies of the ultrastructure and permeability of the blood-brain barrier in the developing corpus callosum in postnatal rat brain using electron dense tracers." *J. Anat.* 184 (Pt 2): 227-37.

Yang, Z., D. Schryvers, et al. (2006). "Demonstration of lanthanum in liver cells by energy-dispersive X-ray spectroscopy, electron energy loss spectroscopy and high-resolution transmission electron microscopy." *J. Microsc.* 223(Pt 2): 133-9.

Yoshikumi, Y., H. Ohno, et al. (2008). "Up-regulation of JAM-1 in AR42J cells treated with activin A and betacellulin and the diabetic regenerating islets." *Endocr. J.* 55(4): 757-65.

Yue, L., J. B. Peng, et al. (2001). "CaT1 manifests the pore properties of the calcium-release-activated calcium channel." *Nature* 410(6829): 705-9.

Index

A

Abraham, 38, 47
absorption, 18, 21, 36, 38, 44, 50, 60
acid, 19, 20, 57
ACTH, 48
adenine, 37
adenosine, 21
adhesion, 4, 48, 52, 55
adjustment, 56
airway epithelial cells, 56
airways, ix, 21
amiloride, 50
amino acids, 13
antioxidant, 24
apoptosis, 19, 23
arginine, 56
arrest, 39
arterial hypertension, 55
arterioles, 14
artery, 55
asthma, ix, 3, 21
astrocytes, 13
ATP, 29
atria, 57
autoimmune diseases, ix, 3
autopsy, 38, 40
avoidance, 30, 46

B

background, 10
bacterial infection, 20
barium, 18, 59
barriers, 25, 45, 55, 60
basal lamina, 13, 15, 22
basement membrane, 23, 33
BBB, 13, 14, 15, 17
bile, 19, 20, 38, 47, 52
bile acids, 20
bilirubin, 19
bioavailability, 36, 57
blood plasma, 24
blood stream, 15
blood-brain barrier, 13, 50, 57, 61
body fluid, 15
bone, 35, 36, 38, 44, 46, 47, 49
brain, ix, 13, 14, 15, 16, 30, 39, 40, 47, 53, 55, 61
breakdown, 27, 50, 54
bronchial epithelial cells, 56
bronchiectasis, 21

C

Ca^{2+}, 1, 4, 9, 21, 29, 30, 31, 39, 45, 46, 47, 50, 52, 53, 57, 58, 59, 60
cadmium, 23

calcification, 35, 37, 56
calcium, ix, 30, 31, 36, 37, 44, 46, 48, 49, 50, 52, 55, 57, 58, 60, 61
calcium carbonate, 36, 37, 46, 49
cancer, ix, 3
capillary, 14
capsule, 31, 58
carcinoma, 18
cardiac muscle, 29, 54
cation, 58
cell culture, 7, 9, 10, 39
cell line, 4, 48, 50, 52, 57
cell lines, 4, 48, 50, 57
cell surface, 31, 33
central nervous system, 16, 38
cerebrospinal fluid, 15, 55
China, 37
cholestasis, 20, 58
choroid, 15, 16, 61
chromium, 23, 55
chronic kidney failure, 48
chronic renal failure, 35, 38, 44, 49, 51, 56
cilia, 16, 21
clinical trials, 37
CO_2, 13
cognitive function, 39
colitis, 19, 49, 53, 54, 58
collagen, 10
colon, 18, 19
color, iv
communication, 16
complexity, 3, 8
composition, 15, 21, 24, 53
conductance, ix, 10, 12
connective tissue, 22, 57
constipation, 36
contamination, 40, 47
contracture, 29
controlled trials, 55
corpus callosum, 61
correlation, x
cortex, 22
cost, 36, 37, 45
CSF, 16, 55
culture, 26

cystic fibrosis, 21, 48, 50, 51, 60
cystitis, 27
cytochrome, 18
cytokines, 14, 20, 23, 56, 59
cytoskeleton, 54

D

damages, iv
decay, 57
defence, 24
deficiency, 23
degenerate, 25
degradation, 38
dementia, 38
dephosphorylation, 56
deposition, 38, 40, 47, 53
deprivation, 55
detachment, 26
detection, 1, 39
developing brain, 15
diabetes, 17
dialysis, 35, 36, 38, 43, 46, 47, 51, 55
diarrhea, 36
diet, 36
diffusion, 11, 13, 16, 20, 25, 53
diploid, 22
discs, 4
dislocation, 44
disposition, 57
distilled water, 7
dorsal horn, 30, 57
Drosophila, 30, 32, 43
drugs, 18, 20, 37
duodenum, 19
dynamics, 58
dyspepsia, 36

E

edema, 14
egg, 24
electrical resistance, ix, 1
electrolyte, 12

electromagnetic, 23, 60
electron, 1, 2, 8, 17, 21, 33, 45, 56, 57, 59, 61
electron microscopy, 1, 17, 21
encephalomyelitis, 15, 50
encephalopathy, 36
endocrine, 37, 50
endocrine glands, 37
endothelial cells, 13, 14, 15, 16, 17, 22, 23, 48
endothelium, 14, 24, 55
endotoxemia, 20
end-stage renal disease, x, 35, 40, 45, 46, 49
enzymes, 24
ependymal, 13, 15, 16
ependymal cell, 13, 15, 16
epidermis, 26
epididymis, ix, 24, 43, 53
epithelia, ix, 1, 11, 45, 46, 47
epithelial cells, 16, 17, 18, 20, 21, 23, 25, 26, 32, 52
epithelium, 17, 18, 19, 20, 21, 25, 26, 27, 43, 46, 50, 51, 52, 53, 54, 55, 56, 58, 60
esophagus, 18
estrogen, 20
ethanol, 8, 19
etiology, 18
excretion, 19
exposure, 39, 49, 60
extracellular matrix, 23

F

filament, 23
filtration, 16
fixation, 9
flatulence, 36
fluid, 16, 20, 24, 45
fluorescence, 45

G

gadolinium, 59
gallbladder, 54
gastric mucosa, 18, 54
gastrointestinal tract, 17, 38
gene expression, 38, 44
gland, 15, 31
glial cells, 15, 16, 17
glucose, 13, 32
glucose-induced insulin secretion, 32
glutathione, 24
goblet cells, 18

H

hallucinations, 38
haploid, 23
harmful effects, 33
health effects, 59
hemodialysis, 39, 43, 49, 51, 58
hepatocytes, 33, 52, 57
hepatotoxicity, 44
histology, 19, 39
homeostasis, 13, 15
human subjects, 57
hydrocephalus, 15, 55
hyperbilirubinemia, 20
hyperparathyroidism, 35
hyperphosphatemia, x, 30, 36, 37, 40, 45, 49
hypertension, 14, 25, 36
hypertonic saline, 47, 48, 60
hypotension, 36
hypothermia, 14
hypothesis, 22

I

ileum, 1, 19, 54
illumination, 26
images, 1
immune system, 23
impregnation, 56
in vivo, 21, 39, 50, 54, 56
incisor, 44
infants, 51
inflammatory bowel disease, 19
inflammatory cells, 5

Index

inhibition, 14, 31
insects, 53
intercellular contacts, 4
interstitial cystitis, 27
intestine, ix, 19, 36, 54
intravenously, 38
invertebrates, 14, 16
ion channels, 3
ion transport, 60
ions, x, 1, 2, 8, 13, 24, 29, 54, 59
iris, 56
irradiation, 26
ischemia, 14, 17, 52, 57
isotope, ix, 7

J

Japan, 37

K

kidney, 32, 35, 37, 39, 43, 45, 48, 51, 58
kidney failure, 45
kidneys, 35

L

lanthanide, 47
leakage, 14
learning, 30, 46
learning task, 30
lens, 17
lipid metabolism, 26
liver, ix, 19, 20, 38, 39, 51, 57, 58, 61
liver cells, 61
localization, 49
long-term memory, 30
lumen, 14, 20, 21, 23, 24
lung function, 48
lying, 22
lymph, 38
lymph node, 38
lymphocytes, 22

M

macromolecules, 22, 25
macrophages, 22, 33
management, 36, 43, 52, 58
mannitol, ix, 11, 14, 21, 47, 55
mast cells, 27
matrix, 23, 58
maturation process, 24
media, 57
medulla, 22
meiosis, 23
membranes, 1, 11, 25, 30
memory, 30, 59
memory retrieval, 59
mesangial cells, 53
meta-analysis, 37
metabolic acidosis, 36
metabolism, 38, 58
metabolites, 13
mice, 27, 60
microscope, 2, 8
microscopy, ix, 7, 49
migration, 22, 23, 51
miniature, 30
molecular weight, 7
molecules, ix, 2, 7, 11, 13, 15, 16, 22, 24, 25, 59, 60
monolayer, 11, 18
morphology, 2, 9, 18
mRNA, 37, 56
mucosa, 19, 59
mucus, 21, 47
mutant, 24, 43, 54
mutation, 58

N

NaCl, 8, 9, 21
nausea, 36
necrosis, 52
nerve, 1, 16, 30, 46, 54, 57, 59
nervous system, 13, 15, 16, 53
neurons, 30, 32, 43, 57

neutrophils, 51
Norway, 53
nutrients, 56

O

obstruction, 20
optic nerve, 56
organ, 15, 39
organelles, 29, 30, 38
osmolality, 56
osteodystrophy, 39
osteomalacia, 37, 38

P

pachytene, 23
pain, 39
palate, 18
parallel, 59
parathyroid, 31, 35, 37, 49, 50
parathyroid hormone, 35, 37, 49
parenchyma, 14
pathogenesis, 19, 20, 47
pathophysiology, 54, 56, 58
pathways, 4, 12, 21, 25, 48, 53
performance, 39
perfusion, 26
permeability, ix, 1, 3, 4, 7, 8, 10, 11, 14, 16, 17, 18, 19, 20, 21, 25, 26, 31, 38, 44, 45, 49, 50, 51, 52, 53, 54, 55, 56, 57, 58, 59, 60, 61
permeation, 58
permission, iv
pertussis, 31
phosphates, 33
phosphorus, 33, 35, 48, 49, 54, 56
phosphorylation, 4, 56
photobleaching, 45
physiology, 54, 56
pigs, 24, 60
pineal gland, 15
pituitary gland, 31
placenta, 25, 52

placental barrier, 25, 51
plasma membrane, 3, 25, 48
plexus, 15, 16, 61
polarity, 3, 4, 48
polymer, 48
portal vein, 38
potassium, 23, 59
precipitation, 1, 8, 9, 58
probe, ix
pro-inflammatory, 46
properties, 2, 12, 15, 31, 32, 61
prosthesis, 18
protease inhibitors, 23
proteases, 23
protein family, 4
proteins, 3, 4, 14, 15, 16, 23, 48, 50, 54
puberty, 23
pyrimidine, 61

R

radio, 7
radius, ix, 9, 11
receptors, 3
recommendations, iv
redistribution, 29
regression, 51
remodelling, 4
renal failure, 38, 43, 53
renal osteodystrophy, 36, 49
repair, 14
resistance, 2, 10, 46
resolution, 61
reticulum, 32
retina, 17, 45, 56
retinal disease, 17
rights, iv
room temperature, 8

S

salts, 45, 49
scattering, 1
secrete, 20

secretion, 15, 21, 26, 31, 59
selectivity, 3, 12, 46
seminiferous tubules, 22, 44
senses, 15
serine, 56
Sertoli cells, 23
serum, 19, 20, 38, 49, 54, 57, 58
sheep, 43
signalling, 59
skeletal muscle, 60
skin, 2, 11, 40, 48, 52
small intestine, 18, 19, 49, 54
sodium, 7, 21, 29, 57, 58, 60
space, ix, 1, 3, 8, 9, 18, 25, 58
spectroscopy, 33, 61
sperm, 24
spermatogenesis, 51, 58
spinal cord, 30
stimulus, 59
stomach, 18
storage, 26, 47
sulfuric acid, 19
Sun, 11, 23, 59
survival, 51
suspensions, 48
Sweden, xii
syndrome, 26, 47

T

teeth, 17
temperature, 10
terminals, 30, 46
testicle, 60
testing, 14
therapeutic agents, 14
therapy, 39, 43, 49, 50, 55
threonine, 56
thymus, 22, 57
thyroid, 37
tissue, 4, 10, 13, 14, 16, 18, 22, 26, 35, 39, 45, 47, 53, 55, 56
TNF, 20
TNF-α, 20
tonic, 31

toxic effect, 36, 38
toxic substances, 15
toxicity, 36, 39
toxin, 31
trace elements, 40
trachea, 33, 57
transmission, ix, 8, 61
transmission electron microscopy, ix, 61
transport, 11, 13, 15, 16, 20, 21, 38, 44, 45, 60, 61
trial, 30, 46, 48
triggers, 15, 48, 57
tumor, 20
tumor necrosis factor, 20
turnover, 36
tyrosine, 56

U

ultrasound, 25, 51
ultrastructure, 61
uniform, 1, 58
urea, 48, 56
urinary bladder, 2, 26, 46, 52
urine, 27
urothelium, 26, 52

V

vas deferens, 24, 31, 50
ventricle, 45
venules, 14, 22
vesicle, 14
vessels, 16
vitamin A, 23
vitamin D, 35, 60
vomiting, 15, 36

W

weakness, 11

X

X-ray, 33, 53, 61

Y

yeast, 18

Z

zinc, 19, 58